TUFTED CARPET

Textile Fibers, Dyes, Finishes, and Processes

TUFTED CARPET

Textile Fibers, Dyes, Finishes, and Processes

by

Von Moody

Manchester, Tennessee

Howard L. Needles, Ph.D.

Pebble Beach, California

WILLIAM ANDREW PUBLISHING
Norwich, New York, U.S.A.

Library of Congress Catalog Card Number:
ISBN: 1-884207-99-5

Published in the United States of America by
William Andrew Publishing
13 Eaton Avenue
Norwich, NY 13815
1-800-932-7045
www.williamandrew.com
www.knovel.com

Transferred to Digital Printing, 2010

Printed and bound in the United Kingdom

NOTICE

Library of Congress Cataloging-in-Publication Data

Moody, Von.
Tufted carpet : textile fibers, dyes, finishes, and processes / Von Moody, Howard Needles.--1st ed.
p. cm.
Includes bibliographical references and index.
ISBN 1-884207-99-5 (alk. paper)
1. Dyes and dyeing--Textile fibers. 2. Textile chemicals. I. Needles, Howard L. II. Title.

TT853.M66 2004
677'.643--dc22

2004002375

DEDICATION

This book is dedicated to my wife, Vickie,
and my mother, Theresa.

Preface

Tufted carpet is made by inserting tufts of yarn through a fabric. It is manufactured by machines that use needles to insert the pile tufts through the fabric backing. Carpeting is available in numerous pile cuts and lengths. Colors and surface textures can be varied, and designs can be printed on them. In North America, about 90% of all carpet is tufted.

The yarns used in tufted carpet are primarily synthetic. Some 97% of all carpet is produced with synthetic fiber, accounting for some 3.5 million pounds of plastics used in the United States alone in 2001. The prominent synthetic fibers and their approximate percent of production are as follows:

- Nylon (59%)
- Polyolefin (polypropylene) (33%)
- Polyester (7%)
- Other fibers (1%)

While the production percentages between these synthetic fiber types do change from year to year, synthetic fibers are firmly established as the fiber of choice for carpets based on cost, efficiency of production, performance, and esthetics.

The tufted carpet industry is a very large industry, both worldwide and in North America. The people who work in this industry are professionals spanning a variety functions, bringing with them a variety of skills and backgrounds. They are chemists, engineers, designers, processors, machinery and equipment producers and operators, marketing and management personnel, and bulk carpet purchasers, including facilities managers and retail store buyers. This book was written as an intermediate book for all carpet industry professionals. Its aim is to serve experienced professionals with a brief, but comprehensive reference for review of industry practices. It aims to serve newcomers to the industry by providing a solid introduction to tufted carpet fundamentals and to serve all industry professionals as a communication tool.

Part 1 of the book, "Introduction to Carpet Fiber," covers the fundamentals of carpet fiber: theory and formation in Chapter 1; identification and characterization in Chapter 2; structural, physical, chemical, and end-use properties in Chapter 3.

The main fibers used in the carpet market today are solution-dyed nylon bulk continuous filament (BCF) and nylon staple. Multi-coloration of the face is typical. Commercial end use represented about 47% of the carpet market in the United States in 2001. Most commercial carpet is made from BCF. Residential carpet, representing about 53% of United States carpet production in 2001, is dominantly staple fiber. Usually, it is dyed to a solid color. Almost 64% of all residential styles are dyed after the carpet is formed. Over 35% of residential carpet is manufactured from yarn that has been dyed. The remaining 1%, for the residential market, is solution-dyed nylon (BCF). This segment may grow when color stability is needed and from special marketing that the fiber companies and carpet manufacturers may do.

Part 2 of the book, "Carpet Construction," covers carpet construction methods: yarn formation in Chapter 4; primary and secondary backing construction in Chapter 5; carpet construction in Chapter 6. There are many methods of making carpet. Originally, carpets were crafted by hand. Simple looms followed and brought about the development of mechanized looms, which revolutionized weaving in the late eighteenth century. The tufting process is an outgrowth of the 1930s chenille bedspread industry. As the machinery for chenille developed, its products expanded to include mats, rugs, and carpets. Today, tufted carpet is produced on machines that use needles to insert parallel rows of tufting into the carpet backing. Machine variations in the movements of the needles and auxiliary tools, such as knives, produce the different styles of loop pile. The most popular commercial style is level loop pile with a low pile height. In the residential market, cut pile with higher pile heights accounts for 66.6% of all styles. Loop pile makes up 25% of the styles and cut/loop makes up the remaining 8.4%.

Part 3 of the book, "Coatings, Raw Materials, and Their Processes," covers carpet system coatings, raw materials, and their processing: latex coatings in Chapter 7; polyurethane coating in Chapter 8; cushions and pads in Chapter 9; polyvinyl chloride plastisol coating in Chapter 10; hot melt coating in Chapter 11; extrusion coating in Chapter 12; carpet tile coatings and reinforcements in Chapter 13. Basic carpet qualities of dimensional stability, adhesion, moisture resistance, fuzzing, aging, flammability, sound insulation, strength, and so on depend on the coatings, processes, and cushions and pads that comprise the entire manufactured carpet system. The industry has developed many reliable practices, detailed in the Part 3 chapters, to ensure these fundamental quality features. Additionally, these elements remain of interest in research for improved carpeting and manufacturing methods.

Part 4 of the book, "Carpet Enhancers," covers colors, decoration, and stain and microbial protection: antimicrobial agents in Chapter 14; color, dyes, dyeing, printing in Chapter 15; stain blockers in chapter 16. Carpet end users are most acquainted with these advanced features, which have their own processing methods and conditions necessary for advertising and warranty claims.

Part 5 of the book, "Performance, Cleaning, and Recycling," covers selected standards, equipment, and processing: carpet performance standards and tests in Chapter 17; maintenance and cleaning in Chapter 18; recycling in Chapter 19. These are important after-market issues and consumer concerns of importance to all professionals in the carpet industry.

We would like to thank the tufted carpet industry for the career opportunities we have had to contribute to this creative, scientifically-challenging and -rewarding, and economically-important enterprise. We hope that our efforts with this book convey our interests, knowledge, and enthusiasm to our many coworkers and that they will be better able to perform in their own jobs because of it.

Von Moody
Howard Needles

January 2004

Contents

Part 1: Introduction to Carpet Fiber

Part 4: Carpet Enhancers

Part 1

Introduction to Carpet Fiber

1

Fiber Theory and Formation

The word *textile* was originally used to define a woven fabric and the processes involved in weaving. Over the years, the term has taken on broad connotations, including the following:

1. Staple filaments and fibers for use in yarn; or preparation of woven, knitted, tufted, or non-woven fabrics.
2. Yarns made from natural or man-made fibers.
3. Fabrics and other products made from fibers or yarns.
4. Apparel or other articles fabricated from the above that retain the flexibility and drape of the original fabrics.

This broad definition generally covers all of the products produced by the textile industry intended for intermediate structures or final products.

Textile fabrics are planar structures produced by interlacing or entangling yarns or fibers in some manner. In turn, *textile yarns* are continuous strands made up of textile fibers, the basic physical structures or elements that make up textile products. Each individual fiber is made up of millions of individual long molecular chains of discrete chemical structure. The arrangement and orientation of these molecules within the individual fiber, as well as the gross cross-section and shape of the fiber (morphology), affects fiber properties; but, by far, the molecular structure of the long molecular chains determines its basic physical and chemical nature. Usually, the polymeric molecular chains found in fibers have a definite

1.2.1 Primary Properties

Fiber Length to Width Ratio. Fibrous materials must have sufficient length so that they can be made into twisted yarns. In addition, the width of the fiber (the diameter of the cross-section) must be much less than the overall length of the fiber; usually the diameter should be 1/100 of the length. The fiber may be "infinitely" long, as found with continuous filament fibers, or as short as 0.5 inches (1.3 cm), as found in staple fibers. Most natural fibers are staple fibers, whereas man-made fibers come in either staple or filament form, depending on processing prior to yarn formation.

Fiber Uniformity. Fibers suitable for processing into yarns and fabrics must be fairly uniform in shape and size. Without sufficient uniformity of dimensions and properties in a given set of fibers to be twisted into yarn, the actual formation of the yarn may be impossible or the resulting yarn may be weak, rough, irregular in size and shape, and unsuitable for textile usage. Natural fibers must be sorted and graded to assure fiber uniformity, whereas synthetic fibers may be "tailored" by cutting into appropriate uniform lengths to give a proper degree of fiber uniformity.

Fiber Strength and Flexibility. A fiber or yarn made from the fiber must possess sufficient strength to be processed into a textile fabric or other textile article. Following fabrication into a textile article, the resulting textile must have sufficient strength to provide adequate durability during end-use. Many experts consider single fiber strength of 5 grams per denier to be necessary for a fiber to be suitable in most textile applications, although certain fibers with strengths as low as 1.0 gram per denier have been found suitable for some applications.

Thestrength of a single fiber is called the *tenacity*, defined as the force per unit linear density necessary to break a known unit of that fiber. The breaking tenacity of a fiber may be expressed in grams per denier (g/d) or grams per tex (g/tex). Both *denier* and *tex* are units of linear density (mass per unit of fiber length) and are defined as the number of grams of fiber measuring 9,000 meters and 1,000 meters, respectively. As a result, the denier of a fiber or yarn will always be nine times the tex of the same fiber. Since tenacities of fibers or yarns are obtained by dividing the force by denier or tex, the tenacity of a fiber in grams per denier will be 1/9 that of the fiber tenacity in grams per tex.

As a result of the adaptation of the International System of Units (SI), the appropriate length unit for breaking tenacity becomes kilometer (km) of breaking length or Newton per tex (N/tex) and will be equivalent in value to g/tex.

1

Fiber Theory and Formation

The word *textile* was originally used to define a woven fabric and the processes involved in weaving. Over the years, the term has taken on broad connotations, including the following:

1. Staple filaments and fibers for use in yarn; or preparation of woven, knitted, tufted, or non-woven fabrics.
2. Yarns made from natural or man-made fibers.
3. Fabrics and other products made from fibers or yarns.
4. Apparel or other articles fabricated from the above that retain the flexibility and drape of the original fabrics.

This broad definition generally covers all of the products produced by the textile industry intended for intermediate structures or final products.

Textile fabrics are planar structures produced by interlacing or entangling yarns or fibers in some manner. In turn, *textile yarns* are continuous strands made up of textile fibers, the basic physical structures or elements that make up textile products. Each individual fiber is made up of millions of individual long molecular chains of discrete chemical structure. The arrangement and orientation of these molecules within the individual fiber, as well as the gross cross-section and shape of the fiber (morphology), affects fiber properties; but, by far, the molecular structure of the long molecular chains determines its basic physical and chemical nature. Usually, the polymeric molecular chains found in fibers have a definite chemical sequence, which repeats itself along the length of the molecule. The total number of units which repeat in a chain (n) varies from a few units to several hundred and is referred to as the *degree of polymerization* (DP) for molecules within that fiber.

1.1 FIBER CLASSIFICATION

Textile fibers are normally broken down into two main classes, natural and man-made fibers (Fig. 1.1). All fibers that come from natural sources (animals, plants, etc.) and do not require fiber formation or reformation are classed as *natural fibers*. Natural fibers include the protein fibers such as wool and silk, the cellulose fibers such as cotton and linen, and the mineral fiber asbestos. *Man-made fibers* are fibers in which either the basic chemical units have been formed by chemical synthesis followed by fiber formation or the polymers from natural sources have been dissolved and regenerated after passage through a spinneret to form fibers. Fibers made by chemical synthesis are often called *synthetic fibers*, while fibers regenerated from natural polymer sources are called *regenerated fibers* or *natural polymer fibers*. In other words, all synthetic fibers and regenerated fibers are man-made fibers, since man is involved in the actual fiber formation process. In contrast, fibers from natural sources are provided by nature in ready-made form.

The synthetic man-made fibers include the polyamides (nylon), polyesters, acrylics, and polyolefins. Figure 1.1 shows a classification chart for the major fibers.

Another method of classifying fibers is by chemical structure without regard to the origin of the fiber and its starting materials. In this manner, all fibers of similar chemical structure are classed together. The natural man-made fiber classification given in Fig. 1.1 does this to a certain extent. In this way, all fibers having the basic cellulosic unit in their structures are grouped together rather than separated into natural and man-made fibers. An outline for the arrangement of fibers by chemical class follows:

- Wool
- Nylon 6 and nylon 6,6
- Polyester
- Acrylic
- Polyolefin

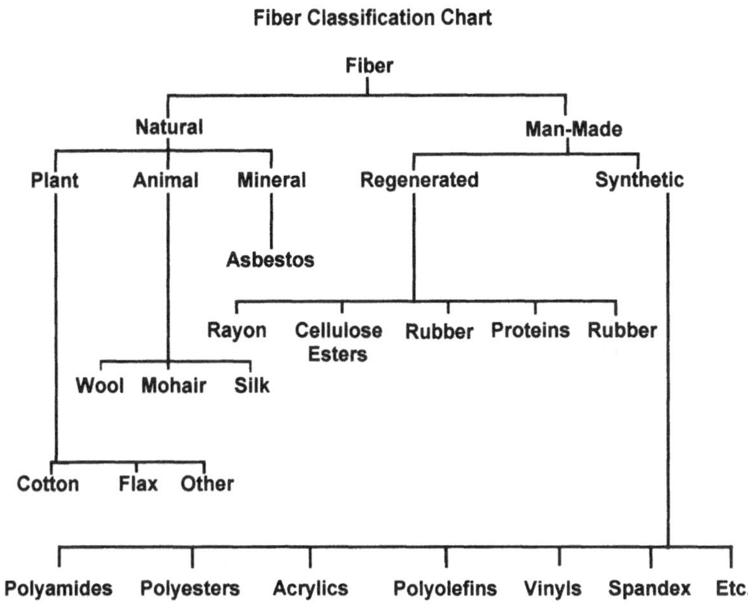

Figure 1.1. Classification of natural and man-made fibers.

1.2 FIBER PROPERTIES

There are several primary properties necessary for a polymeric material to make an adequate fiber:

1. Fiber length to width ratio
2. Fiber uniformity
3. Fiber strength and flexibility
4. Fiber extensibility and elasticity
5. Fiber cohesiveness

Certain other fiber properties increase its value and desirability in its intended end-use, but are not necessary properties essential to make a fiber. Such secondary properties include moisture absorption characteristics, fiber resiliency, abrasion resistance, density, luster, chemical resistance, thermal characteristics, and flammability. A more detailed description of both primary and secondary properties follows in Secs. 1.2.1 and 1.2.2.

1.2.1 Primary Properties

Fiber Length to Width Ratio. Fibrous materials must have sufficient length so that they can be made into twisted yarns. In addition, the width of the fiber (the diameter of the cross-section) must be much less than the overall length of the fiber; usually the diameter should be 1/100 of the length. The fiber may be "infinitely" long, as found with continuous filament fibers, or as short as 0.5 inches (1.3 cm), as found in staple fibers. Most natural fibers are staple fibers, whereas man-made fibers come in either staple or filament form, depending on processing prior to yarn formation.

Fiber Uniformity. Fibers suitable for processing into yarns and fabrics must be fairly uniform in shape and size. Without sufficient uniformity of dimensions and properties in a given set of fibers to be twisted into yarn, the actual formation of the yarn may be impossible or the resulting yarn may be weak, rough, irregular in size and shape, and unsuitable for textile usage. Natural fibers must be sorted and graded to assure fiber uniformity, whereas synthetic fibers may be "tailored" by cutting into appropriate uniform lengths to give a proper degree of fiber uniformity.

Fiber Strength and Flexibility. A fiber or yarn made from the fiber must possess sufficient strength to be processed into a textile fabric or other textile article. Following fabrication into a textile article, the resulting textile must have sufficient strength to provide adequate durability during end-use. Many experts consider single fiber strength of 5 grams per denier to be necessary for a fiber to be suitable in most textile applications, although certain fibers with strengths as low as 1.0 gram per denier have been found suitable for some applications.

Thestrength of a single fiber is called the *tenacity*, defined as the force per unit linear density necessary to break a known unit of that fiber. The breaking tenacity of a fiber may be expressed in grams per denier (g/d) or grams per tex (g/tex). Both *denier* and *tex* are units of linear density (mass per unit of fiber length) and are defined as the number of grams of fiber measuring 9,000 meters and 1,000 meters, respectively. As a result, the denier of a fiber or yarn will always be nine times the tex of the same fiber. Since tenacities of fibers or yarns are obtained by dividing the force by denier or tex, the tenacity of a fiber in grams per denier will be 1/9 that of the fiber tenacity in grams per tex.

As a result of the adaptation of the International System of Units (SI), the appropriate length unit for breaking tenacity becomes kilometer (km) of breaking length or Newton per tex (N/tex) and will be equivalent in value to g/tex.

The strength of a fiber, yarn, or fabric can be expressed in terms of force per unit area, and when expressed in this way, the term is *tensile strength*. The most common unit used in the past for tensile strength has been pounds force per square inch or grams force per square centimeter. In SI units, the pounds force per square inch × 6.895 will become kilopascals (kPa) and grams force per square centimeter × 9.807 will become megapascals (MPa).

A fiber must be sufficiently flexible to go through repeated bendings without significant strength deterioration or breakage of the fiber. Without adequate flexibility, it would be impossible to convert fibers into yarns and fabrics, since flexing and bending of the individual fibers is a necessary part of this conversion. In addition, individual fibers in a textile will be subjected to considerable bending and flexing during end use.

Fiber Extensibility and Elasticity. An individual fiber must be able to undergo slight extensions in length (less than 5%) without breakage of the fiber. At the same time, the fiber must be able to almost completely recover following slight fiber deformation. In other words, the extension deformation of the fiber must be nearly elastic. These properties are important because the individual fibers in textiles are often subjected to sudden stresses, and the textile must be able to give and recover without significant overall deformation of the textile.

Fiber Cohesiveness. Fibers must be capable of adhering to one another when spun into a yarn. The cohesiveness of the fiber may be due to the shape and contour of the individual fibers or the nature of the surface of the fibers. In addition, long-filament fibers by virtue of their length can be twisted together to give stability without true cohesion between fibers. Often the term "spinning quality" is used to state the overall attractiveness of fibers for one another.

1.2.2 Secondary Properties

Moisture Absorption and Desorption. Most fibers tend to absorb moisture (water vapor) when in contact with the atmosphere. The amount of water absorbed by the textile fiber depends on the chemical and physical structure and properties of the fiber, as well as the temperature and humidity of the surroundings. The percentage absorption of water vapor by a fiber is often expressed as its *moisture regain*. The regain is determined by weighing a dry fiber, then placing it in a room set to standard temperature and humidity [21° ± 1°C and 65% relative humidity (RH)] are commonly used). From these measurements, the percentage moisture regain of the fiber is determined.

Eq. (1.1) Percentage regain =

$$\frac{\text{Conditioned weight - Dry weight}}{\text{Dry weight}} \times 100\%$$

Percentage moisture content of a fiber is the percentage of the total weight of the fiber, which is due to the moisture present, and is obtained from the following formula.

Eq. (1.2) Percentage moisture content =

$$\frac{\text{Conditioned weight - Dry weight}}{\text{Conditioned weight}} \times 100\%$$

The percentage moisture content will always be the smaller of the two values.

Fibers vary greatly in their regain, with hydrophobic (water-repelling) fibers having regains near zero and hydrophilic (water-seeking) fibers like cotton, rayon, and wool having regains as high as 15% at 21°C and 65% RH. The ability of fibers to regain large amounts of water affects the basic properties of the fiber in end use. Absorbent fibers are able to absorb large amounts of water before they feel wet, an important factor where absorption of perspiration is necessary. Fibers with high regains will be easier to process, finish, and dye in aqueous solutions, but will dry more slowly. The low regain found for many man-made fibers makes them quick drying, a distinct advantage in certain applications. Fibers with high regains are often desirable because they provide a "breathable" fabric which can conduct moisture from the body to the outside atmosphere readily, due to their favorable moisture absorption-desorption properties. The tensile properties of fibers as well as their dimensional properties are affected by moisture.

Fiber Resiliency and Abrasion Resistance. The ability of a fiber to absorb shock and recover from deformation and to be generally resistant to abrasion forces is important to its end-use and wear characteristics. In consumer use, fibers in fabrics are often placed under stress through compression, bending, and twisting (torsion) forces under a variety of temperature and humidity conditions. If the fibers within the fabric possess good elastic recovery properties from such deformative actions, the fiber has good resiliency and better overall appearance in end use. For example, wool shows poor wrinkle recovery under hot moist conditions, whereas polyester exhibits good recovery from deformation as a result of its high resiliency. Resistance of a fiber to damage when mobile forces or stresses come in contact with fiber structures is referred to as *abrasion resistance*. If a fiber is able to effectively absorb and dissipate these forces without damage, the fiber will show good abrasion resistance. The toughness and hardness of the fiber is related to its chemical and physical structure and the morphology of the fiber, and will influence the abrasion of the fiber. A rigid, brittle fiber such as glass, which is unable to dissipate the forces of abrasive action, results in fiber damage and breakage, whereas a tough, but more plastic, fiber such as polyester shows better resistance to abrasion forces. Finishes can affect fiber properties including resiliency and abrasion resistance.

Luster. *Luster* refers to the degree of light that is reflected from the surface of a fiber or the degree of gloss or sheen that the fiber possesses. The inherent chemical and physical structure and shape of the fiber can affect the relative luster of the fiber. With natural fibers, the luster of the fiber is dependent on the morphological form that nature gives the fiber, although the relative luster can be changed by chemical and/or physical treatment of the fiber in processes, such as mercerization of cotton. Man-made fibers can vary in luster from bright to dull depending on the amount of delusterant added to the fiber. Delusterants such as titanium dioxide tend to scatter and absorb light, thereby making the fiber appear duller. The desirability of luster for a given fiber application will vary and is often dependent on the intended end use of the fiber in a fabric and on current fashion trends.

Resistance to Chemicals in the Environment. A textile fiber, to be useful, must have reasonable resistance to the chemicals it comes in contact with in its environment during use and maintenance. It should have resistance to oxidation by oxygen and other gases in the air, particularly in the presence of light, and be resistant to attack by microorganisms and other biological agents. Many fibers undergo light-induced reactions, and fibers from natural sources are susceptible to biological attack, but such deficiencies can be minimized by treatment with appropriate finishes. Textile fibers come in contact with a large range of chemical agents during laundering and dry cleaning and must be resistant to attack under such conditions.

Density. The density of a fiber is related to its inherent chemical structure and the packing of the molecular chains within that structure. The density of a fiber will have a noticeable effect on its aesthetic appeal and its usefulness in given applications. For example, glass and silk fabrics of the same denier would have noticeable differences in weight due to their broad differences in density. Fishnets of polypropylene fibers are of great utility because their density is less than that of water. Densities are usually expressed in units of grams per cubic centimeter, but in SI units will be expressed as kilograms per cubic meter, which gives a value one thousand times larger.

Thermal and Flammability Characteristics. Fibers used in textiles must be resistant to wet and dry heat, must not ignite readily when coming in contact with a flame, and ideally should self-extinguish when the flame is removed. Heat stability is particularly important to a fiber during dyeing and finishing of the textile, and during cleaning and general maintenance by the consumer. Textile fibers for the most part are made up of organic polymeric materials containing carbon and burn on ignition from a flame or other propagating source. The chemical structure of a fiber establishes its overall flammability characteristics, and appropriate textile finishes can reduce the degree of flammability. A number of federal, state, and local statutes eliminate the most dangerous flammable fabrics from the marketplace.

1.2.3 Primary Fiber Properties from an Engineering Perspective

The textile and polymer engineer must consider a number of criteria essential for formation, fabrication, and assembly of fibers into textile substrates. Often, the criteria used will be similar to those set forth concerning end-use properties. Ideally, a textile fiber should have the following properties:

1. A melting and/or decomposition point above 220°C.
2. A tensile strength of 5 g/denier or greater.
3. Elongation at break above 10% with reversible elongation up to 5% strain.
4. A moisture absorptivity of 2%–5% moisture uptake.
5. Combined moisture regain and air entrapment capability.
6. High abrasion resistance.
7. Resistance to attack, swelling, or solution in solvents, acids, and bases.
8. Self-extinguishes when removed from a flame.

1.3 FIBER FORMATION AND MORPHOLOGY

Fiber morphology refers to the form and structure of a fiber, including the molecular arrangement of individual molecules and groups of molecules within the fiber. Most fibers are organic materials derived from carbon combined with other atoms such as oxygen, nitrogen, and halogens. The basic building blocks that organic materials form as covalently bonded organic compounds are called *monomers*. Covalent bonds involve the sharing of electrons between adjacent atoms within the monomer. The structure of the monomer is determined by the type, location, and nature of bonding of atoms within the monomer and by the nature of covalent bonding between atoms. Monomers react or condense to form long-chain molecules called *polymers* made up of a given number (n) of monomer units, which are the basic building unit of fibers. On formation into fibers and orientation by natural or mechanical means, the polymeric molecules possess ordered crystalline and non-ordered amorphous areas, depending on the nature of the polymer and the relative packing of molecules within the fiber. For a monomer (A), the sequence of events to fiber formation and orientation would appear as shown in Fig. 1.2.

Polymers with repeating units of the same monomer $(A)_n$ would be referred to as *homopolymers*. If a second unit (B) is introduced into the basic structure, copolymers are formed with structures as outlined in Fig. 1.3.

1.3.1 Polymer Formation

Synthetic polymers used to form fibers are often classified on the basis of their mechanism of polymerization—step growth (condensation) or chain growth (addition) polymerization. *Step growth* polymerization involves multifunctional monomers, which undergo successive condensation with a second monomer or with itself to form a *dimer*, which in turn condenses with another dimer to form a *tetramer*, etc., usually with loss of a small molecule such as water. *Chain growth* involves the instantaneous growth of a long molecular chain from unsaturated monomer units, followed by initiation of a second chain, etc. The two methods are outlined below schematically.

Eq. (1.3) $$\text{Step growth}: nA \rightarrow \frac{n}{2} AA \rightarrow \frac{n}{4} AAAA \rightarrow \ldots$$

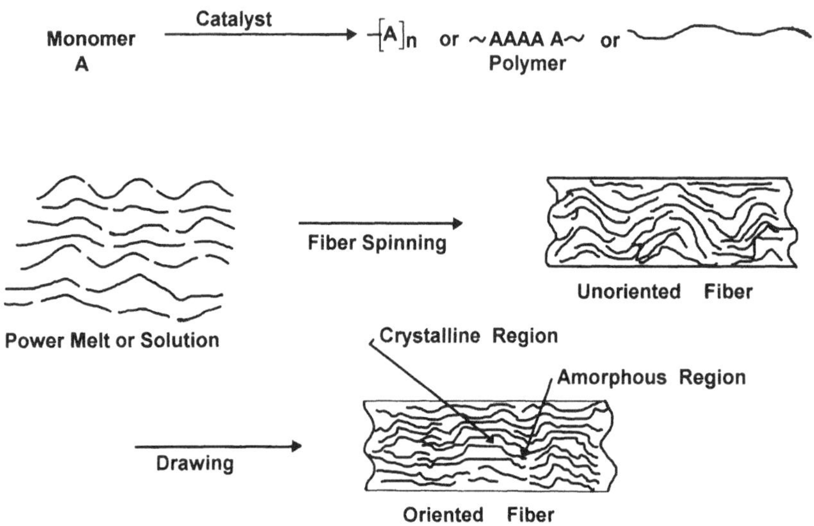

Figure 1.2. Polymerization sequence and fiber formation.

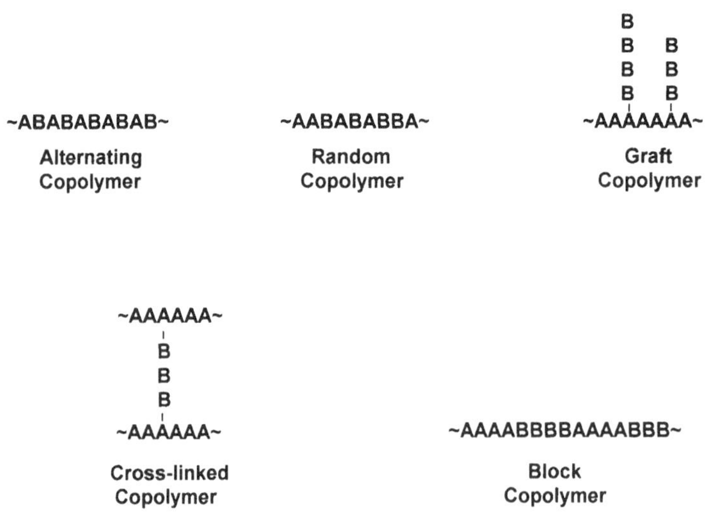

Figure 1.3. Copolymer structures.

Eq. (1.4) Chain growth : $nA \rightarrow (A)_n nA \rightarrow (A)_n nA \rightarrow (A)_n$

The average number of monomer repeating units in a polymer chain (n) is often referred to as the degree of polymerization (DP). The DP must exceed an average twenty units in most cases to give a polymer sufficient molecular size to have desirable fiber-forming properties. The overall breadth of distribution of molecular chain lengths in the polymer will affect the ultimate properties of the fibers, with wide polymer size distributions leading to an overall reduction of fiber properties. Although the polymers from natural fibers and regenerated natural fibers do not undergo polymerization by the mechanisms found for synthetic fibers, most natural polymers have characteristic repeating units and high degrees of polymerization and are related to step growth polymers. Basic polymeric structures for the "major fibers" are given in Fig. 1.4.

1.3.2 Fiber Spinning

Although natural fibers come in a morphological form determined by nature, regenerated and synthetic man-made fibers can be "tailor-made" depending on the shape and dimensions of the orifice (spinning jet) that the polymer is forced through to form the fiber. There are several methods used to spin a fiber from its polymer, including melt, dry, wet, emulsion, and suspension spinning.

Melt spinning is the least complex of the methods. The polymer from which the fiber is made is melted and then forced through a spinneret and into air which causes solidification and fiber formation.

Dry and wet spinning processes involve dissolving the fiber-forming polymer in an appropriate solvent, followed by passing a concentrated solution (20%–50% polymer) through the spinneret and into dry air to evaporate the solvent in the case of dry spinning, or into a coagulating bath to cause precipitation, or regeneration of the polymer in fiber form in the case of wet spinning. There is a net contraction of the spinning solution on loss of solvent. If a skin of polymer is formed on the fiber followed by diffusion of the remainder of the solvent from the core of the forming fiber, the cross section of the fiber as it contracts may collapse to form an irregular popcorn-like cross section.

Emulsion spinning is used only for those fiber-forming polymers that are insoluble. Polymer is mixed with a surface-active agent (detergent), and possibly a solvent, and then mixed at high speed with water to form an emulsion of the polymer. The polymer is passed through the spinneret and into a coagulating bath to form the fiber.

In suspension spinning, the polymer is swollen and suspended in a swelling solvent. The swollen, suspended polymer is forced through the spinneret into dry hot air to drive off the solvent, or into a wet non-solvent bath to cause the fiber to form through coagulation.

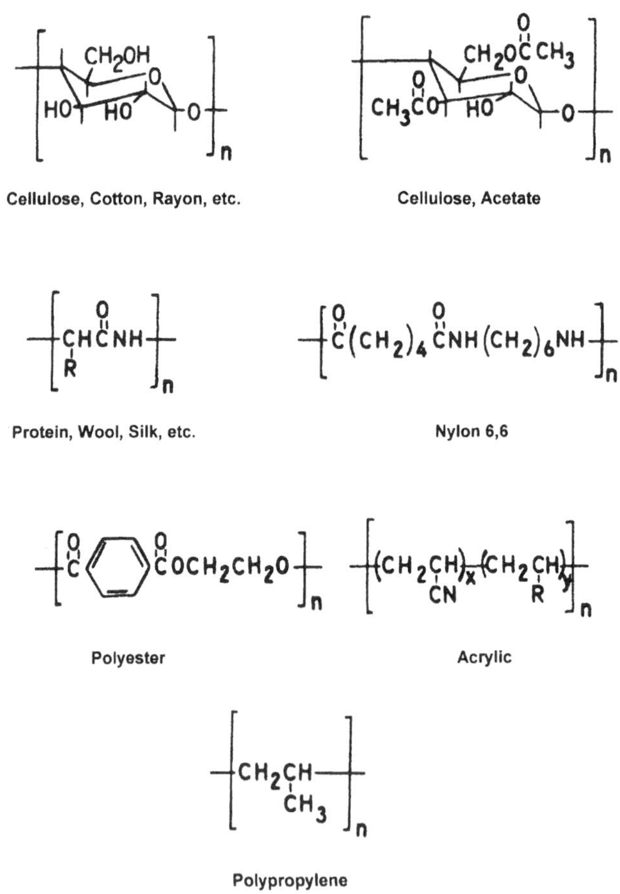

Figure 1.4. Basic polymeric structures for major fibers.

The spinning process can be divided into three steps:

1. Flow of spinning fluid within and through the spinneret under high stress and sheer.
2. Exit of fluid from the spinneret with relief of stress and an increase in volume (ballooning of flow).
3. Elongation of the fluid jet as it is subjected to tensile force as it cools and solidifies with orientation of molecular structure within the fiber.

Common cross sections of man-made fibers include round, trilobal, pentalobal, dog-bone, and crescent shapes. When two polymers are used in fiber formation as in bicomponent or biconstituent fibers, the two components can be arranged in a matrix, side-by-side, or sheath-core configuration. Round cross sections are also found where skin formation has caused fiber contraction and puckering (as with rayon) has occurred, or where the spinneret shape has provided a hollow fiber. Complex fiber cross-sectional shapes with special properties are also used (Fig. 1.5).

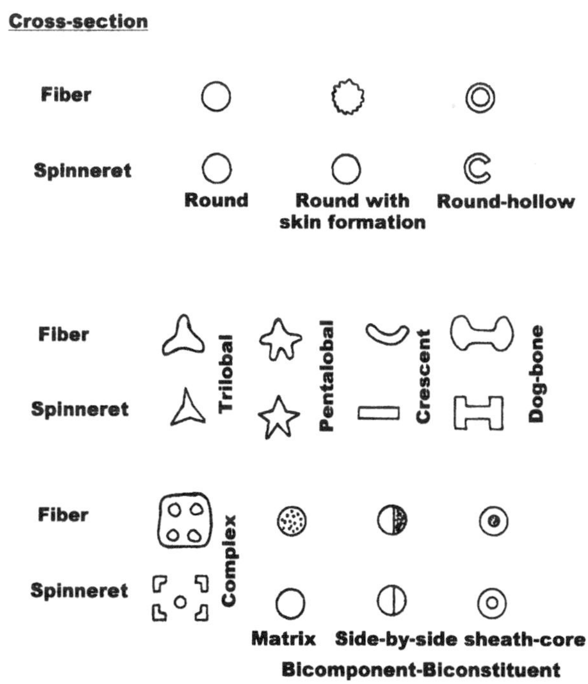

Figure 1.5. Fiber cross section.

1.3.3 Fiber Drawing and Morphology

On drawing and orientation, the man-made fibers become smaller in diameter and more crystalline, and imperfections in the fiber morphology are improved somewhat. Side-by-side bicomponent or biconstituent fibers on drawing become wavy and bulky.

In natural fibers, the orientation of the molecules within the fiber is determined by the biological source during the growth and maturity process of the fiber.

The form and structure of polymer molecules in relation to each other within the fiber will depend on the relative alignment of the molecules in relationship to one another. Those areas where the polymer chains are closely aligned and packed together are crystalline areas within the fiber, whereas those areas where there is essentially no molecular alignment are referred to as amorphous areas. Dyes and finishes can penetrate the amorphous portion of the fiber, but not the ordered crystalline portion.

A number of theories exist concerning the arrangement of crystalline and amorphous areas within a fiber. Individual crystalline areas in a fiber are often referred to as *microfibrils*. Microfibrils can associate into larger crystalline groups, which are called *fibrils* or *micelles*. Microfibrils are 30–100 Å (10^{-10} meters) in length, whereas fibrils and micelles are usually 200–600 Å in length. This compares to the individual molecular chains, which vary from 300 to 1,500 Å in length and which are usually part of both crystalline and amorphous areas of the polymer and, therefore, give continuity and association of the various crystalline and amorphous areas within the fiber. A number of theories have been developed to explain the interconnection of crystalline and amorphous areas in the fiber and include such concepts as fringed micelles or fringed fibrils, molecular chain folding, and extended chain concepts. The amorphous areas within a fiber will be relatively loosely packed and associated with each other, and spaces or voids will appear due to discontinuities within the structure. Figure 1.6 outlines the various aspects of internal fiber morphology with regard to polymer chains.

The forces that keep crystalline areas together within a fiber include chemical bonds (covalent, ionic) as well as secondary bonds (hydrogen bonds, van der Waals forces, dipole-dipole interactions). *Covalent bonds* result from sharing of electrons between atoms, such as found in carbon-carbon, carbon-oxygen, and carbon-nitrogen bonding, within organic compounds. Covalent bonds joining adjacent polymer chains are referred to as cross links. *Ionic bonding* occurs when molecules donate or accept electrons from each other, as when a metal salt reacts with acid side chains on a polymer within a fiber. Chemical bonds are much stronger than secondary bonds formed between polymer chains, but the total associative force between polymer chains can be large since a very large number of such bonds may occur between adjacent polymer chains. *Hydrogen bonds* are the strongest of the secondary bonds and occur between electropositive hydrogen atoms and electronegative atoms such as oxygen, nitrogen, and halogens on opposing polymer chains. Nylon, protein, and cellulosic fibers are capable of extensive hydrogen bonding. Van der Waals interactions between polymer chains occur when clouds of electrons from each chain come in close proximity, thereby promoting a small attractive force between chains. The more extended the cloud of electrons, the stronger the van der Waals interaction will be. Covalently bonded materials will show some uneven distribution of electron density over the molecule due to the differing electronegativity of the atoms and electron distribution over the molecule to form dipoles. *Dipoles* on adjacent polymer chains of opposite charge and close proximity are attracted to each other and promote secondary bonding. When a synthetic fiber is stretched or drawn, the molecules in most cases will orient themselves in crystalline areas parallel to the fiber axis, although crystalline areas in some chain-folded polymers such as polypropylene can be aligned vertically to the fiber axis. The degree of crystallinity will be affected by the total forces available for chain interaction, the distance between parallel chains, and the similarity and uniformity of adjacent chains. The structure and arrangement of individual polymer chains also affects the morphology of the fiber. Also, *cis-trans* configurations or optical isomers of polymers can have very different physical and chemical properties.

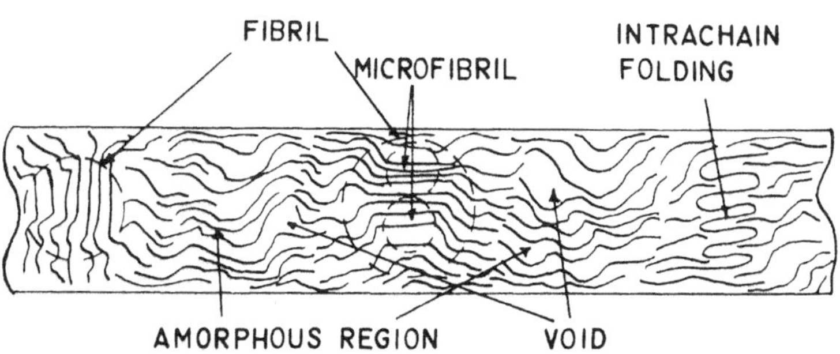

Figure 1.6. Aspects of internal fiber morphology.

1.3.4 Bulking and Texturizing

Thermoplastic man-made fibers can be permanently heat set after drawing and orientation. The fiber will possess structural integrity and will not shrink up to that setting temperature. Also, thermoplastic fibers or yarns from these fibers can be texturized to give three-dimensional loft and bulkiness:

1. Through fiber deformation and setting at or near their softening temperature.
2. Through air entanglement.
3. Through differential setting within fibers or yarns (Table 1.1).

Schematic representations of these methods are given in Fig. 1.7.

Heat-Setting Techniques. Six heat-setting techniques are currently in use.

False Twist Heat-Setting Technique. The false twist heat-setting technique is extremely rapid, inexpensive, and the most widely used. The filament fiber tow is brought in contact with a high-speed spindle running vertically to the moving tow. This action results in a high twist in the tow up to the spindle. The twisted tow is heated near its softening point before passing the spindle, then cooled, and untwisted to give a wavy, bulky yarn.

Table 1.1. Texturizing Methods

Heat-setting Techniques	Air Entanglement	Differential Setting
False twist Knife edge Stuffer box Gear crimping Autotwist Knit-de-knit	Air jet	Bicomponent-biconstituent Fiber orientation Heat shrinkage of thermoplastic fibers in a blend

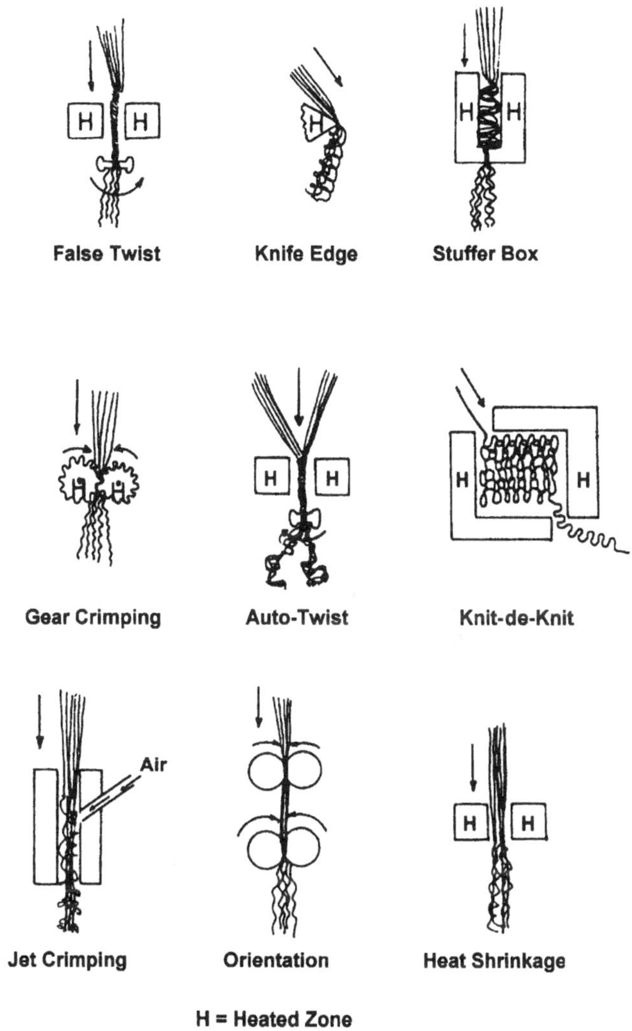

Figure 1.7. Texturizing methods.

Knife-Edge Texturizing. In knife-edge texturizing, filament tows or yarns under tension are passed over a heated knife edge. The fibers near the knife edge are changed in overall orientation in relation to the unheated yarns or portion of the filaments away from the knife edge, thereby causing bulking of the yarn.

Stuffer Box Texturizing. In stuffer box texturizing, the filament tow is fed into a heated box, causing the tow to double up against itself. On removal, the cooled tow retains the zigzag configuration caused by the process.

Gear Crimping. In gear crimping, the tow is passed between heated intermeshing gears. On cooling, the fibers retain the shape induced by the heated gears.

Autotwisting. In autotwisting, two tows or yarns are twisted together and then heat set. On untwisting, the yarns have equal, but opposite twists which cause a spiral bulking of the yarn.

Knit-de-Knit. In the knit-de-knit process, a yarn is fill-knitted, heat set, cooled, and de-knitted to give a bulked yarn retaining the shape and curvature of the knit.

Air Entanglement. In air entanglement texturizing, a fiber tow is loosely fed into and through a restricted space and a high-speed air jet is impinged on the fibers at a 45° angle. The loose fibers within the tow are looped to give a texturized effect.

Differential Setting. Heat shrinkage techniques cause a bulking of fiber tows containing different fibers through heating one component of the blend sufficiently to cause heat shrinkage of the fiber and compaction, contraction, and bulking. Side-by-side bicomponent and biconstituent fibers recover different degrees on each side from fiber stretching causing a waving, crimping, or bulking of the fiber.

1.3.5 Staple Formation

Continuous filaments can be cut into staple by wet or dry cutting techniques. In *wet cutting*, the wet-spun fiber is cut to uniform lengths right after spinning, while *dry cutting* involves partial cutting, debonding, and shuffling of the dry tow to form a sliver.

Before the filament or staple is used in yarn spinning, spin finishes are added to give lubricity and antistatic characteristics to the fibers and to provide a greater degree of fiber cohesiveness. The finishes are usually mixtures including such materials as fatty acid esters, mineral oils, synthetic esters, silicones, cationic amines, phosphate esters, emulsifiers, and/or nonionic surfactants. Spin finishes are formulated to be oxidation resistant, to be easily removed by scouring, to give a controlled viscosity, to be stable to corrosion, to resist odor and color formation, and to be nonvolatile and readily emulsifiable.

1.4 STRUCTURE-PROPERTY RELATIONSHIPS

The basic chemical and morphological structure of polymers in a fiber determine the fundamental properties of a fabric made from that fiber. Although physical and chemical treatments and changes in yarn and fabric formation parameters can alter the fabric properties to some degree, the basic properties of the fabric result from physical and chemical properties inherent to the structure of the polymer making up the fiber. From these basic properties, the end-use characteristics of the fiber are determined. To that end, Ch. 3, "Major Fibers and Their Properties," describes the various textile fibers in terms of their basic structural properties, their physical and chemical properties, and finally the end-use characteristics inherent to constructions made from the fiber.

Initially, the name and general information for a given fiber is set forth followed by an outline of the structural properties, including information about the chemical structure of the polymer, the degree of polymerization, and the arrangement of molecular chains within the fiber. Physical properties include mechanical (tensile) and environmental properties of the fiber, whereas the effect of common chemicals and chemically-inducted processes on the fibers are listed under chemical properties. The end-use properties are then listed and include properties coming inherently from the structural, physical, and chemical properties of the fiber, as well as end-use properties that involve evaluation of performance, subjective aspects, and aesthetics of the fabrics. Where possible, Ch. 3, "Major Fibers and Their Properties," presents the interrelationships of these properties.

2

Fiber Identification and Characterization

Fibers make up the face, and sometimes the backing of the carpet. The characteristics and qualities of the fiber are a major determinant of the performance of the carpet.

2.1 FIBER IDENTIFICATION

Several methods are used to identify fibers and to differentiate them from one another. The most common methods include microscopic examination, solubility tests, heating and burning characteristics, density or specific gravity, and staining techniques.

2.1.1 Microscopic Identification

Examination of longitudinal and cross-sectional views of a fiber at 100 to 500 magnifications gives detailed information about the surface morphology of the fiber. Positive identification of many natural fibers is possible using the microscope, but positive identification of man-made fibers is more difficult due to their similarity in appearance and due to the fact that spinning techniques and spinneret shape can radically alter the gross morphological structure of the fiber.

2.1.2 Solubility

The chemical structure of polymers in a fiber determines the fiber's basic solubility characteristics, and the effect of solvents on fibers can aid in the general fiber classification. Various classification schemes involving solubility have been developed to separate and identify fibers.

2.1.3 Heating and Burning Characteristics

The reaction of fibers to heat from an open flame is a useful guide in the identification of fibers. When thermoplastic fibers are brought close to a flame, they melt, fuse, and shrink, whereas nonthermoplastic fibers brown, char, or are unaffected by the flame. On contact with an open flame, fibers of organic polymers ignite and burn. The nature of the burning reaction is characteristic of the chemical structure of the fiber. On removal from the flame, fibers either self-extinguish or continue to burn. The odor of gases coming from the decomposing fibers and the nature of any residual ash are characteristic of the fibrous polymer being burned.

2.1.4 Density or Specific Gravity

Fiber density may be used as an aid in fiber identification. Fiber density may be determined by using a series of solvent mixtures of varying density or specific gravity. If the specific gravity of the fiber is greater than that of the liquid, the fiber specimen sinks in the liquid. Conversely, if the specific gravity of the fiber is less than that of the liquid, the fiber specimen floats. Thereby, an approximate determination of fiber density may be made.

2.1.5 Staining

Fibers have differing dyeing characteristics and affinities dependent on the chemical and morphological structure of the fiber. Prepared dye mixtures containing dyes of differing affinities for various fiber types have been used extensively as identification stains for undyed fabrics. Since some fiber types may dye to similar shades with these dye mixtures, two or more stains usually must be used to confirm the fiber content. Staining is effective only for previously undyed fibers or for fibers where the dye is stripped from the fiber prior to staining.

2.2 STRUCTURAL, PHYSICAL, AND CHEMICAL CHARACTERIZATION

A number of methods are available for characterization of the structural, physical, and chemical properties of fibers. The major methods available are outlined in this chapter, including a brief description of each method and the nature of characterization that the method provides.

2.2.1 Optical and Electron Microscopy

Optical microscopy (OM) has been used for many years as a reliable method to determine the gross morphology of a fiber in longitudinal, as well as cross-sectional views. Mounting the fiber on a slide wetted with a liquid of appropriate refractive properties has been used to minimize light scattering effects. The presence of gross morphological characteristics such as fiber shape and size and the nature of the surface can be readily detected. Magnifications as high as 1,500X are possible, although less depth of field exists at higher magnifications. Scanning electron microscopy (SEM) can be used to view the morphology of fibers with good depth of field and resolution at magnifications up to 10,000X. In scanning electron microscopy, the fiber must first be coated with a thin film of a conducting metal such as silver or gold. The mounted specimen then is scanned with an electron beam, and back-scattered particles emitted from the fiber surface are detected and analyzed to form an image of the fiber. Transmission electron microscopy (TEM) is more specialized and more difficult to perform than SEM. It measures the net density of electrons passing through the thin cross sections of metal-coated fibers and provides a method to look at their micro-morphologies.

2.2.2 Elemental and End-Group Analysis

The qualitative and quantitative analysis of the chemical elements and groups in a fiber may aid in identification and characterization of a fiber. Care must be taken in analysis of such data, since the presence of dyes or finishes on the fibers may affect the nature and content of elements and end groups found in a given fiber. Gravimetric and instrumental chemical methods are available for analysis of specific elements or groups of elements in fibers. Specific chemical analyses of functional groups and end groups in organic polymers that make up fibers may be carried out. For example, analyses of amino acids in protein fibers, amino groups in polyamides and proteins, and acid groups in polyamides and polyesters aid in structure determination, molecular characterization, and identification of fibers.

2.2.3 Infrared Spectroscopy

Infrared spectroscopy is a valuable tool in determination of functional groups within a fiber. Functional groups in a polymer absorb infrared energy at wavelengths characteristic of the particular group and lead to changes in the vibrational modes within the functional group. As a result of the infrared absorption characteristics of the fiber, specific functional groups can be identified. Infrared spectroscopy of fibers can be carried out on the finely divided fiber segments pressed in a salt pellet, or through the use of reflectance techniques. Functional groups in dyes and finishes also can be detected by this technique.

2.2.4 Ultraviolet-Visible Spectroscopy

The ultraviolet-visible spectra of fibers, dyes, and finishes can provide clues concerning the structure of these materials, as well as show the nature of electronic transitions that occur within the material as light is absorbed at various wavelengths by unsaturated groups giving an electronically-excited molecule. The absorbed energy is either harmlessly dissipated as heat, fluorescence, or phosphorescence, or causes chemical reactions to occur that modify the chemical structure of the fiber. Ultraviolet-visible spectra can be measured for a material either in solution or by reflectance. Reflectance spectra are particularly useful in color measurement and assessment of color differences in dyed and bleached fibers.

2.2.5 Nuclear Magnetic Resonance Spectroscopy

Nuclear magnetic resonance (NMR) spectroscopy measures the relative magnitude and direction (moment) of spin orientation of the nucleus of the individual atoms within a polymer from a fiber in solution in a high-intensity magnetic field. The degree of shift of spins within the magnetic field and the signal splitting characteristics of individual atoms such as hydrogen or carbon within the molecule are dependent on the location and nature of the groups surrounding each atom. In this way, the "average" structure of long polymeric chains can be determined. Line width from NMR spectra also can provide information concerning the relationship of crystalline and amorphous areas within the polymer.

2.2.6 X-Ray Diffraction

X-rays, diffracted from or reflected off crystalline or semicrystalline polymeric materials, give patterns related to the crystalline and amorphous areas within a fiber. The size and shape of individual crystalline and amorphous sites within the fiber are reflected in the geometry and sharpness of the x-ray diffraction pattern and provide an insight into the internal structure of the polymeric chains.

2.2.7 Thermal Analysis

Physical and chemical changes in fibers may be investigated by measuring changes in selected properties as small samples of fiber are heated at a steady rate over a given temperature range in an inert atmosphere such as nitrogen. There are four thermal characterization methods.

1. Differential thermal analysis (DTA)
2. Differential scanning calorimetry (DSC)
3. Thermal gravimetric analysis (TGA)
4. Thermal mechanical analysis (TMA)

In DTA, small changes in temperature (ΔT) in the fiber sample compared to a reference are detected and recorded as the sample is heated. The changes in temperature (ΔT) are directly related to physical and chemical events occurring within the fiber as it is heated. These events include changes in crystallinity and crystal structure, loss of water, solvents and volatile materials, and melting and decomposition of the fiber. Differential scanning calorimetry is similar to DTA, but measures changes in heat content (ΔH) rather than temperature (ΔT) as the fiber is heated; it provides quantitative data on the thermodynamic processes involved. In an inert gas such as nitrogen, most processes are endothermic (heat absorbing). If DTA or DSC is carried out in air with oxygen, data may be obtained related to the combustion characteristics of the fiber, and fiber decomposition becomes exothermic (heat generating). Thermal gravimetric analysis measures changes in mass (ΔM) of a sample as the temperature is raised at a uniform rate. It provides information concerning loss of volatile materials, the rate and mode of decomposition of the fiber, and the effect of finishes on fiber decomposition. Thermal mechanical analysis measures changes in a specific mechanical property as the temperature of the fiber is raised at a uniform rate. A number of specialized mechanical devices have been developed to measure mechanical changes in fibers, including hardness and flow under stress.

2.2.8 Molecular Weight Determination

Molecular weight determination methods provide information concerning the average size and distribution of individual polymer molecules making up a fiber. Molecular weights enable one to calculate the length of the average repeating unit within the polymer chain, better known as the DP. The distribution of polymer chain lengths within the fiber provides information concerning selected polymer properties.

The major molecular weight determination methods include number average molecular weights ($\overline{M}_n$), determined by end-group analysis, osmometry, cryoscopy, and ebullioscopy; weight average molecular weights ($\overline{M}_w$), determined by light scattering and ultracentrifugation; and viscosity molecular weights ($\overline{M}_v$), determined by the flow rate of polymer solutions. Since each method measures the average molecular weight of the polymer differently, the molecular weight values obtained will differ depending on the overall number and distribution of polymer chains of varying lengths present in the fiber. The differences in value between $\overline{M}_n$ and $\overline{M}_w$ provide measures of the breadth of distribution of polymers within the fiber. By definition the distribution of molecular weights for a given polymer will always be $\overline{M}_w > \overline{M}_v > \overline{M}_n$.

2.2.9 Mechanical and Tensile Property Measurements

Mechanical and tensile measurements for fibers include tenacity or tensile strength, elongation at break, recovery from limited elongation, stiffness (relative force required to bend the fiber), and recovery from bending. The tensile properties of individual fibers or yarns are usually measured on a tensile testing machine such as an Instron®, which subjects fibers or yarns of a given length to a constant rate of force or loading. The force necessary to break the fiber or yarn, or *tenacity*, is commonly given in grams per denier (g/d) or grams per tex (g/tex), or as kilometer breaking length in the SI system. The *elongation to break* of a fiber is a measure of the ultimate degree of extension that a fiber can withstand before breaking. The degree of recovery of a fiber from a given elongation is a measure of the *resiliency* of the fiber to small deformation forces. The stiffness or bendability of a fiber is related to the overall chemical structure of the macromolecules making up the fiber, the forces between adjacent polymer chains, and the degree of crystallinity of the fiber. Mechanical and tensile property measurements can provide valuable insights into the structure of a fiber and its projected performance in end use.

2.2.10 Specific Gravity

The *specific gravity* of a fiber is a measure of its density in relation to the density of the same volume of water, and provides a method to relate the mass per unit volume of a given fiber to that of other fibers. The specific gravity relates in some degree to the nature of molecular packing, crystallinity, and molecular alignment in the fiber. Specific gravity of a fiber will give an idea of the relative weight of fabrics of identical fabric structure, but of differing fiber content. End-use properties such as hand (feel or touch), drapability, and appearance are affected by fiber density.

2.2.11 Environmental Properties

Environmental properties include those physical properties which relate to the environment in which a fiber is found. Moisture regain, solvent solubility, heat conductivity, the physical effect of heat, and electrical properties depend on the environmental conditions surrounding the fiber. The uptake of moisture by a dry fiber at equilibrium will depend on the temperature and relative humidity of the environment. Solvent solubilities of fibers will depend on the solubility parameters of the solvent in relation to fiber structure and crystallinity. Heat conductivity, the physical effect of heating such as melting, softening, and other thermal transitions, and the electrical properties of a fiber depend on the inherent structure of the fiber and the manner in which heat or electrical energy is acted upon by the macromolecules within the fiber. Environmental properties are measured by subjecting the fiber to the appropriate environmental conditions and measuring the property desired under such conditions.

2.2.12 Chemical Properties

The chemical properties of fibers include the effects of chemical agents like acids, bases, oxidizing agents, reducing agents, and biological agents such as molds and mildews on the fiber, and light- and heat-induced chemical changes within the fiber. Acids and bases cause hydrolytic attack of molecular chains within a fiber, whereas oxidizing and reducing agents cause chemical attack of functional groups through oxidation (removal of electrons) or reduction (addition of electrons). Such chemical attack can change the fiber's structure and possibly cleave the molecular chains within the fiber. Biological agents such as moths on wool or mildew on cellulose use the fiber as a nutrient for biological growth and, subsequently, cause damage to the fiber structure.

Sunlight contains ultraviolet, visible, and infrared light energy. This energy can be absorbed at discrete wavelength ranges by fibers depending on their molecular structure. Ultraviolet and visible light absorbed by a fiber will cause excitation of electrons within the structure, raising them to higher energy states. Shorter ultraviolet wavelengths are the most highly energetic and give the most highly excited states. Visible light usually has little effect on the fiber, although its absorption and reflectance of unabsorbed light will determine the color and reflectance characteristics of the fiber. Infrared energy absorbed will increase the vibration of molecules within the fiber and will cause heating. The excited species within the fiber can return to their original (ground) state, through dissipation of the energy as molecular vibrations or heat, without significantly affecting the fiber. Ultraviolet and some visible light absorbed by the fiber, however, can lead to molecular scission within the fiber and cause adverse free radical reactions, which will lead to fiber deterioration.

Heating a fiber to progressively higher temperatures in air will lead to physical as well as chemical changes within the fiber. At sufficiently high temperatures, molecular scission, oxidation, and other complex chemical reactions associated with decomposition of the fiber will occur causing possible discoloration and a severe drop in physical and end-use properties for the fiber.

2.3 END-USE PROPERTY CHARACTERIZATION

End-use property characterization methods often involve use of laboratory techniques which are adapted to simulate actual conditions of average wear on the textile or that can predict performance in end-use. Often quantitative numerical values cannot be listed in comparing the end-use properties of a given textile fiber; nevertheless, relative rankings are possible and can give useful information about the suitability for a specific application of a fabric made from a given fiber type. It must be emphasized that extreme care must be taken in interpreting results from test methods and extrapolating the findings to actual wear and use conditions.

The ultimate properties of fibers in end use do reflect the underlying morphological, physical, and chemical characteristics inherent to the fiber. All major end-use properties and characteristics considered in this handbook are outlined in Secs. 2.3.1 to 2.3.3. End-use methods are usually voluntary or mandatory standards developed by test or trade organizations or by government agencies. Organizations involved in standards development for textile end use include the following:

- American Association of Textile Chemists and Colorists (AATCC)
- American National Standards Institute (ANSI)
- American Society for Testing and Materials (ASTM)
- Consumer Product Safety Commission (CPSC)
- Federal Trade Commission (FTC)
- Society of Dyers and Colorists (SDC)
- International Standards Organization (ISO)

2.3.1 Characteristics Related to Identity, Aesthetics, and Comfort

Fibers are known by common, generic, and trade names. The Textile Fiber Products Identification Act, administered by the Federal Trade Commission, established generic names for all major classes of fibers based on the structure of the fiber. Common natural fibers often are also designated by their variety, type, or country of origin, whereas man-made fibers manufactured by various firms are designated by trade names. Nevertheless, the labeled textile must include the generic name of the fiber(s) and the percentage content of each fiber within the textile substrate. Often trade names are selected which conveys to the consumer a particular "feeling," property, or use for that fiber. Nylon is an example of a trade name (selected by DuPont for their polyamide fiber) which came into such common usage that the Federal Trade Commission (FTC) eventually designated it as the generic name of this fiber class. As new fibers of novel structure are developed and commercialized, the FTC designates new generic names.

A number of fiber end-use properties in textile constructions relate to the aesthetic, tactile, and comfort characteristics of the fiber. Such properties include appearance, luster, hand (feel or touch), drapability, absorbency, overall comfort, crease retention, pilling, and wrinkle resistance. All of these factors are affected to varying extents by the particular properties desired from the textile structure and its intended use. Many of these properties are related to inherent properties of the fibers, which are translated into textile structures prepared for end use.

The overall appearance and luster of a textile can be related to the shape and light absorbing and scattering characteristics of the individual fibers within the structure. The *hand* or *handle* of a textile structure is a complex synthesis of tactile responses by an individual, and is characteristic of the particular fiber or fiber blend and overall structure of the textile substrate. The drapability of textiles is related to the fiber stiffness and bendability within the complex structural matrix making up the textile. The moisture absorbency and comfort of a fiber is related to its chemistry and morphology and to the way it absorbs, interacts with, and conducts moisture. In addition, comfort is related to the yarn and fabric structure into which the individual fibers have been made. Crease retention and wrinkle resistance of a fiber in a textile construction are directly related to the inherent chemical and morphological characteristics of the fiber as they depend on deformation and recovery under dry and moist conditions. The pilling characteristics of a fiber in a textile construction are related to the ease with which individual fibers may be partially pulled from the textile structure and to the tenacity of the individual fibers. Fibers in a loose, open textile structure are readily pulled from the textile. If the fiber is strong, the fiber tangles with other loose fibers and mixes with lint and fiber fragments to form a *pill*. Weaker fibers such as cotton, however, usually break off before pill formation occurs.

2.3.2 Characteristics Related to Durability and Wear

The useful life of a fabric depends on a number of factors, including the strength, stretch, recovery, toughness, and abrasion resistance of the fiber and the tearing and bursting resistance of the fabrics made from that fiber. The composite of these factors coupled with the conditions and type of end use or wear will determine the durability characteristics of a textile structure made from the fiber.

Fibers must be of minimum strength in order to construct textile structures with reasonable wear characteristics. The wear and durability of a fabric will tend to increase with increasing fiber strength. Textile structures made from fibers able to withstand stretching and deformation with good recovery from deformation will have improved durability, particularly when subjected to bursting or tearing stresses. The relative toughness of the fiber also will affect the fabric durability, with tougher fibers giving the best performance. Tough, but resilient, fibers will also be resistant to abrasion or wear by rubbing the fiber surface. Abrasion of a textile structure usually occurs at edges (edge abrasion), on flat surfaces (flat abrasion), or through flexing of the textile structure resulting in interfiber abrasion (flex abrasion).

2.3.3 Physical and Chemical Characteristics and Response of Fiber to Its Environmental Surroundings

The physical and chemical characteristics of a fiber affect a number of important end-use properties: (1) heat (physical and chemical) effect on fibers, including the safe ironing temperature and flammability, (2) wetting of and soil removal from the fiber, including laundering, dry cleaning, and fiber dyeability and fastness, and (3) chemical resistance, including resistance to attack by household chemicals and atmospheric gases, particularly in the presence of sunlight.

Fibers respond to heat in different ways. Thermoplastic fibers such as polyesters soften and eventually melt on heating without extensive decomposition, thereby permitting setting of the softened fiber through stretching and/or bending and subsequent cooling. Other fibers, such as the cellulosic and protein fibers, decompose before melting and, therefore, cannot be set using physical means. The safe ironing temperature of a fabric is determined by the softening and/or decomposition temperature of the fiber and must be significantly below this temperature. At sufficiently high temperatures, fibers are chemically attacked by oxygen in the atmosphere, which accelerates fiber decomposition. If the temperature and heat input is sufficiently high or if a flame is involved, the fiber will ignite and burn and, thereby, decompose at a more rapid rate. On removal from the heat source, some fibers will self-extinguish, whereas others will sustain a flame and continue to burn. The burning characteristics of a fiber depend on its inherent chemical structure and the nature of any finishes or additives present on the fiber.

When soil is removed from a fabric as in laundering or dry cleaning, the individual fibers must be resistant to attack or damage caused by additives such as detergents, the solvent medium used, and mechanical agitation. Fabrics constructed of fibers that swell in water or dry-cleaning solvents can undergo profound dimensional changes on wetting. Also, fibers with surface scales such as wool undergo felting in the presence of moisture and mechanical action.

The dyeability of a fiber is dependent on the chemical and morphological characteristics of the fiber, the ability of the fiber to be effectively wetted and penetrated by the dyeing medium, and the diffusion characteristics of the dye in the fiber. Since most dyeing processes are done in water medium, hydrophilic fibers generally dye more readily than the more hydrophobic fibers. The fastness of the dye on the fiber will be dependent on the nature and order of physical and/or chemical forces holding the dye on the fiber and the effect of environmental factors such as sunlight, household chemicals, and mechanical action (crocking) on the dye-fiber combination.

The chemical resistance of a fiber can have a profound effect on end use. The fibers that are sensitive to chemical attack by household chemicals such as bleach are limited in their end uses. The resistance of fibers to attack by atmospheric gases including oxygen, ozone, and oxides of nitrogen, particularly in the presence of sunlight and moisture, can also be important considerations in certain end uses.

3

Major Fibers and Their Properties

The properties of the major fibers used in carpet manufacture are presented in this chapter. Nylon yarns account for over 70% of the yarns used to form the tufted face of the substrate, with polyester, polypropylene, acrylic, modacrylic, and wool yarns being used to lesser extents. Nylon dominates the tufted carpet market due to its overall toughness and resiliency. Polypropylene is also used both in primary and secondary backing.

3.1 NYLON 6 AND NYLON 6,6 FIBERS

The polyamide fibers include the nylons, 6 and 6,6, and the aramid fibers. Both fiber types are formed from polymers of long-chain polyamides. The nylons generally are tough, strong, durable fibers useful in a wide range of textile applications. The number of carbon atoms in each monomer or comonomer unit is commonly designated for the nylons. Therefore, the nylon with six carbon atoms in the repeating unit would be nylon 6 and the nylon with six carbons in each of the monomer units would be nylon 6,6.

Nylon 6 and nylon 6,6 are very similar in properties and structure (Fig. 3.1) and, therefore, are described together. The major structural difference is due to the placement of the amide groups in a continuous head-to-head arrangement in nylon 6, whereas in nylon 6,6, the amide groups reverse direction each time in a head-to-tail arrangement due to the differences in the monomers and polymerization techniques used.

Nylon 6,6 was developed in the United States, whereas nylon 6 was developed in Europe, and more recently in Japan. The major differences in the fibers are that nylon 6,6 dyes lighter, has a higher melting point, and a slightly harsher hand than nylon 6.

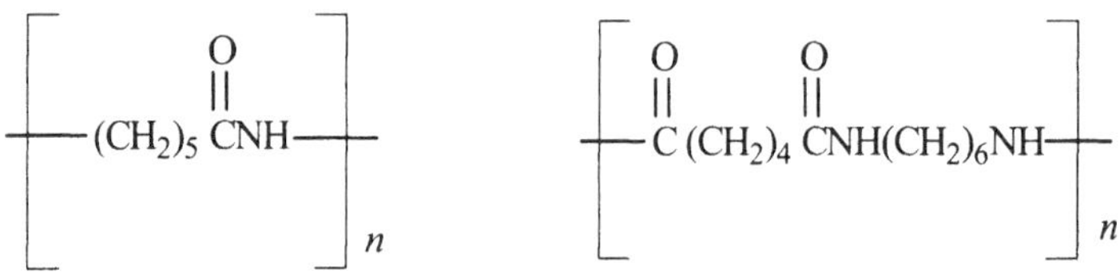

Nylon 6 Nylon 6,6

Figure 3.1. Chemical structures of nylon 6 and nylon 6,6.

3.1.1 Structural Properties

Nylon 6 is produced by ring-opening chain growth polymerization of caprolactam in the presence of water vapor and an acid catalyst at the melt. After removal of water and acid, the nylon 6 is melt spun at 250°–260°C into fibers. Nylon 6,6 is prepared by step growth polymerization of hexamethylene diamine and adipic acid. After drying, the nylon 6,6 is melt spun at 280°–290°C into fibers. Both nylon 6 and 6,6 are drawn to mechanically orient the fibers following spinning.

The degree of polymerization of nylon 6 and 6,6 molecules varies from 100 to 250 units. The polyamide molecular chains lay parallel to one another in a "pleated sheet" structure with strong hydrogen bonding between amide linkages on adjacent molecular chains. The degree of crystallinity of the nylon will depend on the degree of orientation given to the fiber during drawing. Nylon fibers are usually rodlike with a smooth surface or are trilobal in cross section (Figs. 3.2 and 3.3). Multilobal (star) cross sections and other complex cross sections are also found.

3.1.2 Effect of Single-Step Versus Two-Step Production of Nylon

Nylon fibers produced in a single-step process tend to have a more open polymer structure compared to nylon produced in a two-step process. The cause for this difference is the polymer structure is not allowed to relax or condition in the single-step process prior to the typical second step. This relaxation or conditioning step allows the nylon to form more crystalline structure within the polymer matrix. Fibers produced in a single-step method tend to abrade and stain more easily than those manufactured in a two-step process.

High levels of noncrystalline structure create an easily dyeable fiber. This also increases the likelihood of staining and soiling along with poor wear performance. More stain resistant chemical is applied to a fiber of lower crystallinity to achieve the same performance as a fiber of higher crystallinity.

Figure 3.2. Nylon 6,6, round ×1900.

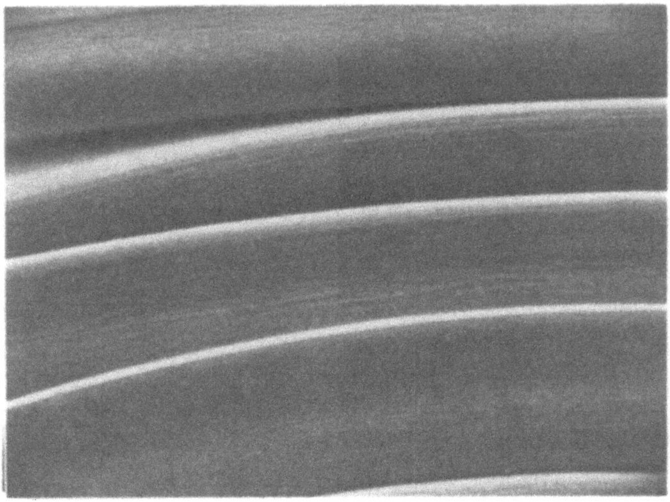

Figure 3.3. Nylon 6,6, trilobal ×1200.

3.1.3 Physical Properties

Nylon 6 and 6,6 fibers are strong, with a dry tenacity of 4–9 g/d (36–81 g/tex) and a wet tenacity of 2.5–8 g/d (23–72 g/tex). These nylons have elongations at break of 15%–50% dry, which increase somewhat on wetting. Recovery from stretch deformation is very good, with 99% recovery from elongations up to 10%. The nylons are stiff fibers with excellent resiliency and recovery from bending deformation. They are of low density, with a specific gravity of 1.14. They are moderately hydrophilic with a moisture regain of 4%–5% under standard conditions and a regain of 9% at 100% RH. Nylon 6 and 6,6 are soluble in hydrogen bond-breaking solvents such as phenols, 90% formic acid, and benzyl alcohol. They have moderate heat conductivity properties and are unaffected by heating below 150°C. The nylons have high resistivity and readily build up static charge.

3.1.4 Chemical Properties

The nylons are fairly resistant to chemical attack. They are attacked by acids, bases, and reducing and oxidizing agents under extreme conditions not found in normal use. They are unaffected by biological agents, but at elevated temperatures or in the presence of sunlight, they will undergo oxidative degradation with yellowing and loss of strength.

3.1.5 Nylon End-Use Properties

Nylon has the following properties:

1. The fiber is tough and has good abrasion resistance.
2. Nylon 6 and nylon 6,6 are extremely strong fibers with excellent recovery and resiliency.
3. It has a low moisture content.
4. The fiber has good resistance to household chemicals, but exhibits poor resistance to attack by sunlight unless treated with antioxidants.
5. Sunlight degrades the polymer over a prolonged period of time.
6. Continuous filaments hide soil better than staple ones.
7. Nylon fibers have high luster unless delustered.
8. Static electricity can easily be generated.
9. Decomposition occurs in strong mineral acids.
10. Soil-hiding properties are changed by the shape of the fiber.
11. Melting occurs at 414°–480°F.
12. Lit cigarettes easily melt it.
13. Its Limiting Oxygen Index (LOI), the amount of oxygen in air necessary to cause combustion, is 20.
14. Nylons melt, drip, and tend to self extinguish on burning.

15. Continuous filaments are typically used in high traffic areas.
16. Fibers have excellent dyeability with excellent color-fastness.
17. Nylon 6 is somewhat deeper dyeing than nylon 6,6.
18. Staining is a problem unless the product is treated with a stain resistant chemical.

3.2 POLYESTER FIBERS

Polyesters are those fibers containing at least 85% of a polymeric ester of a substituted aromatic carboxylic acid including, but not restricted to, terephthalic acid and *p*-hydroxybenzoic acid. The major polyester in commerce is polyethylene terephthalate, an ester formed by step growth polymerization of terephthalic acid and the diol ethylene glycol. Poly-1,4-cyclohexylenedimethylene terephthalate is the polyester of more limited usage and is formed through the step growth polymerization of terephthalic acid with the more complex diol 1,4-cyclohexylenedimethanol. The polyester fibers all have similar properties, are highly resilient and resistant to wrinkling, possess high durability and dimensional stability, and are resistant to chemical and environmental attack.

Polyethylene terephthalate polyester is the leading man-made fiber in production volume and owes its popularity to its versatility alone or as a blended fiber in textile structures. When the term "polyester" is used, it refers to this generic type. It is used extensively in woven and knitted apparel, home furnishings, and industrial applications. Modification of the molecular structure of the fiber through texturizing and or chemical finishing extends its usefulness in various applications.

3.2.1 Structural Properties

Polyethylene terephthalate (Fig. 3.4) is formed through step growth polymerization of terephthalic acid or dimethyl terephthalate with ethylene glycol at 250°–300°C in the presence of a catalyst to a DP of 100–250.

The resultant polymer is isolated by cooling and solidification and dried. Polyester fibers are melt spun from the copolymer at 250°–300°C, followed by fiber orientation, and stretching. The polyester molecular chains are fairly stiff and rigid due to the presence of periodic phenylene groups along the chain. The polyester molecules within the fiber tend to pack lightly and are held together by van der Waals forces. The polyesters are highly crystalline unless comonomers are introduced to disrupt the regularity of the molecular chains. Polyester fibers are usually smooth and rodlike with round or trilobal cross sections (Fig. 3.5).

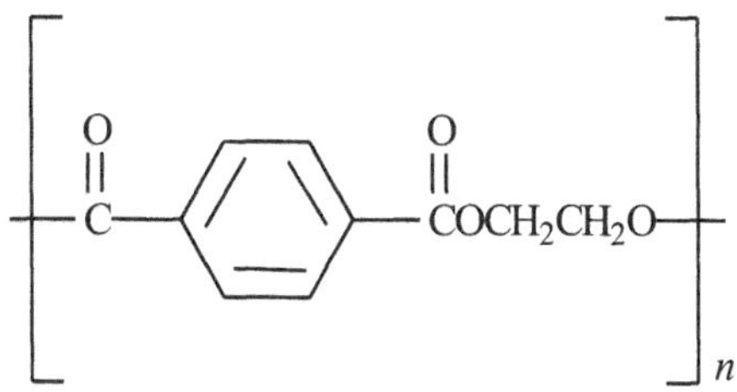

Polyethylene Terephthalate Polyester

Figure 3.4. Structure of polyethylene terephthalate polyester.

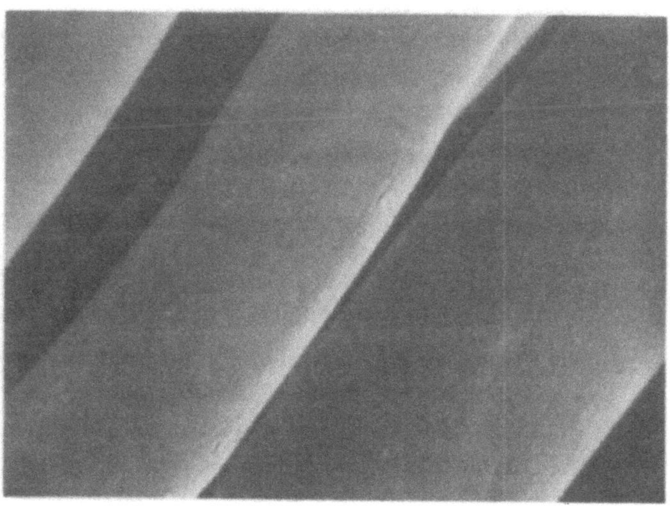

Figure 3.5. Polyester ×1000.

3.2.2 Physical Properties

Polyester from polyethylene terephthalate is an extremely strong fiber with a tenacity of 3–9 g/d (27–81 g/tex). The elongation at break of the fiber varies from 15% to 50% depending on the degree of orientation and nature of crystalline structure within the fiber. The fiber shows moderate (80%–95%) recovery from low elongations (2%–10%). The fiber is relatively stiff and possesses excellent resiliency and recovery from bending deformation. The fiber has a specific gravity of 1.38. The fiber is quite hydrophobic, with a moisture regain of 0.1%–0.4% under standard conditions and 1.0% at 21°C and 100% RH. It is swollen or dissolved by phenols, chloroacetic acid, or certain chlorinated hydrocarbons at elevated temperatures. The fiber exhibits moderate heat conductivity and has high resistivity, leading to extensive static charge buildup. On heating, the fiber softens in the 210°–250°C range with fiber shrinkage and melts at 250°–255°C.

3.2.3 Chemical Properties

Polyethylene terephthalate polyester is highly resistant to chemical attack by acid, bases, oxidizing and reducing agents, and is only attacked by hot concentrated acids and bases. Biological agents do not attack the fiber. On exposure to sunlight, the fiber slowly undergoes oxidative attack without color change with an accompanying slow loss in strength. The fiber melts at about 250°C with only limited decomposition.

3.2.4 Polyester End-Use Properties

Polyester has the following properties:

1. Staple is the most common form because filaments tend to crush and not recover.
2. Polyester possesses good strength and durability characteristics, but exhibits moderate to poor recovery from stretching.
3. Polyester's durability is better than wool.
4. Its abrasion resistance is good.
5. It has a LOI of 21.

6. It is a moderately flammable fiber that burns on contact with a flame, but melts, drips, and shrinks away from the flame.
7. Static build up occurs.
8. The fiber is hydrophobic and nonabsorbent without chemical modification.
9. Due to its hydrophobicity and high crystallinity, polyester is difficult to dye and special dyes and dyeing techniques must be used.
10. When dyed, polyester generally exhibits excellent colorfastness properties.
11. Oily soil is retained unless treated with appropriate soil-release agents.
12. It has excellent resistance to most household chemicals and is resistant to sunlight-induced oxidative damage.
13. It has a bright translucent appearance unless a delusterant has been added to the fiber.
14. Polyester is typically made from recycled materials.
15. Stain resistance is good.

3.3 ACRYLIC FIBERS

The acrylic fibers include acrylic, modacrylic, and other vinyl fibers containing cyanide groups as side chains. Among the major acrylic fibers, acrylonitrile is the comonomer containing a cyanide group. Acrylic fibers are formed from copolymers containing greater than 85% acrylonitrile monomer units, whereas modacrylic fibers contain 35%–85% acrylonitrile units. In general, these fibers possess a warm bulky hand, good resiliency and wrinkle resistance, and overall favorable aesthetic properties.

Acrylic fibers are formed from wet or dry spinning of copolymers. After texturizing, acrylic fibers have a light, bulky, wool-like hand and overall wool-like aesthetics. The fibers are resilient and possess excellent acid and sunlight resistance. Acrylics have been used extensively in applications formerly reserved for wool or other keratin fibers.

3.3.1 Structural Properties

Acrylic fibers are made up of copolymers containing at least 85% acrylonitrile units in combination with one or more comonomers including methyl methacrylate, vinyl acetate, or vinyl pyridine (Fig. 3.6) The copolymer is formed through free radical emulsion polymerization. After precipitation, the copolymer is dried and dissolved in an appropriate organic solvent and wet or dry spun. The degree of polymerization of the copolymers used for fiber formation varies from 150 to 200 units. Pure polyacrylonitrile will form satisfactory fibers. Owing to the extensive tight packing of adjacent molecular chains and the high crystallinity of the fiber, comonomers must be introduced to lower the regularity and crystallinity of the polymer chains to make the fiber more dyeable. Extensive hydrogen bonding occurs between α-hydrogens and the electronegative nitrile groups on adjacent polymer chains, and strong van der Waals interactions further contribute to the packing of the acrylic chains. The periodic comonomer units interfere with this packing and, therefore, decrease the overall crystallinity of acrylic fibers. Acrylic fibers are usually smooth with round or dog-bone cross sections (Fig. 3.7).

3.3.2 Physical Properties

Acrylic fibers are fibers of moderate strength and elongations at break. The tenacity of acrylic fibers varies from 2 to 4 g/d (18–36 g/tex). On wetting, the tenacity drops to 1.5–3 g/d (13–27 g/tex). The elongation at break varies from 20% to 50% for the various acrylic fibers. At 2% elongation, the recovery of the fiber is 99%; however at 5% elongation, the recovery is only 50%–95%. The fiber is moderately stiff and has excellent resiliency and recovery from bending deformation. The fibers have low specific gravities of 1.16–1.18 and low moisture regains of 1.0%–2.5% under standard temperature and humidity conditions. The fiber is soluble in polar aprotic solvents such as *N*, *N*-dimethylformamide. The fiber exhibits good heat and electrical insulation properties. Acrylic fibers do build up moderate static charge and soften at 190°–250°C.

3.3.3 Chemical Properties

Acrylic fibers exhibit good chemical resistance. The fibers are only attacked by concentrated acids and are slowly attacked and hydrolyzed by weak bases. Acrylics are unaffected by oxidizing and reducing agents except for hypochlorite solutions at elevated temperatures. Acrylic fibers are unaffected by biological agents and sunlight. On heating above 200°C, acrylic fibers soften and undergo oxidative attack by a complex mechanism with formation of condensed unsaturated chromophoric (colored) groups in the fiber.

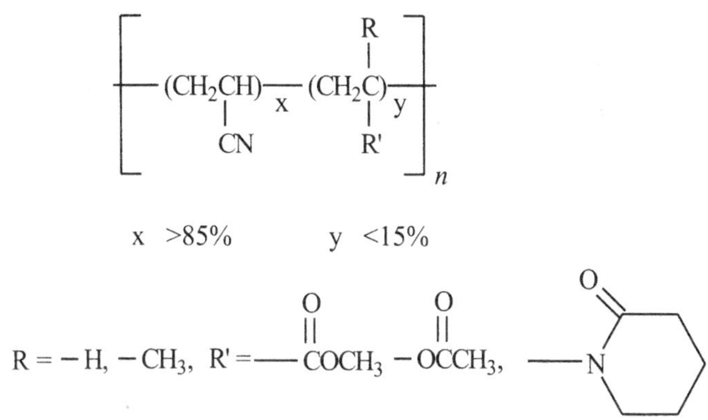

Figure 3.6. Structure of acrylic.

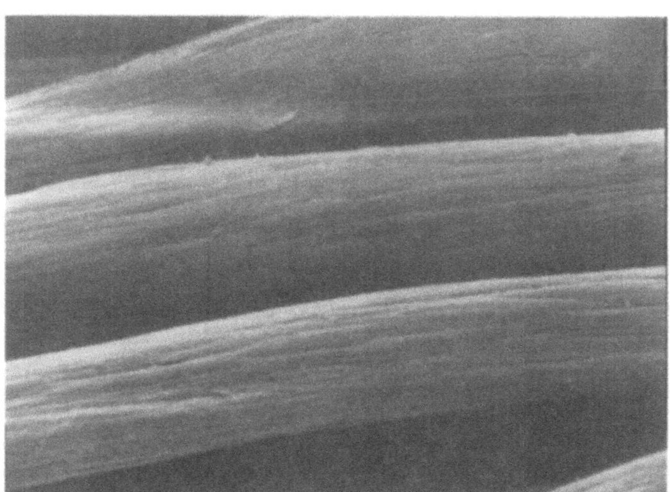

Figure 3.7. Acrylic ×1300.

3.3.4 Acrylic End-Use Properties

Acrylic has the following properties:

1. Acrylic fibers are usually texturized, they have a bulky, wool-like hand and possess a moderate degree of luster.
2. Staple is common due to its wool-like appearance.
3. Crimping and spinning produces a bulky and resilient product like polyester.
4. High bulking does not relate to performance.
5. Yarn weight needs to be approximately the same as wool to perform like wool.
6. They possess fair abrasion and pilling resistance.
7. Bulk and abrasion resistance are similar to wool.
8. Acrylic fiber possesses good resistance to household chemicals and sunlight and is moderately resistant to heat-induced oxidation and discoloration.
9. Owing to the introduction of a comonomer, acrylic fibers are generally dyeable and give fast colors with a wide range of dyes including acid, basic, or disperse dyes.
10. The comonomer present determines the type of dye(s) that may be effectively used.
11. Acrylics have poor appearance retention.
12. There is low static generation.
13. The fibers burn with melting and continue to burn on withdrawal from the flame.
14. The acrylic fibers are moderately flammable with a LOI of 18.
15. Acrylics melt at 420°–490°F.

3.4 MODACRYLIC FIBERS

Modacrylic fibers are formed from copolymers consisting of 35%–85% acrylonitrile and a suitable vinyl comonomer or comonomers such as vinyl chloride, vinylidene chloride, vinyl acetate, vinyl pyrollidone, or methyl acrylate. The modacrylics generally resemble acrylics; have a warm, pleasing hand; and good drapability, resiliency, and wrinkle resistance. They are more heat sensitive, but more flame resistant than acrylics, and have generally been used in specialty applications. Modacrylic fiber exhibits a more thermoplastic character than the related acrylic fibers.

3.4.1 Structural Properties

Modacrylic fibers are wet or dry spun from copolymers of acrylonitrile and an appropriate comonomer or comonomers (Fig. 3.8). The copolymer is formed through free radical chain growth emulsion or solution polymerization to a DP of 150–500. The copolymer is isolated and dissolved in acetone or a similar low-boiling-point solvent, and wet or dry spun to a fiber of round, dog-bone, crescent, or polylobal cross section (Fig. 3.9). The oriented fiber possesses a low crystallinity due to the irregularity and heterogeneity of the copolymer structure. Limited hydrogen bonding and van der Waals interactions are possible due to the limited regularity of adjacent polymer chains.

3.4.2 Physical Properties

The modacrylic fiber is of moderate strength with a dry tenacity of 1.5–3 g/d (14–27 g/tex) and a slightly reduced wet tenacity of 1–2.5 g/d (9–23 g/tex). The fiber has a high elongation at break of 25%–45% and excellent recovery (95%–100%) from low degrees of stretching (<5%). The fibers possess excellent resiliency, moderate stiffness, and have specific gravities of 1.30–1.37. Modacrylic fibers exhibit a wide range of moisture regains, from 0.4% to 4%, depending on the nature and composition of comonomers making up the copolymer. Modacrylic fibers are soluble in ketone solvents such as acetone and in aprotic polar solvents including *N,N*-dimethylformamide. The fibers are good heat and electrical insulators, but tend to build up static charge. The fibers soften in the 135°–160°C range with an accompanying heat shrinkage.

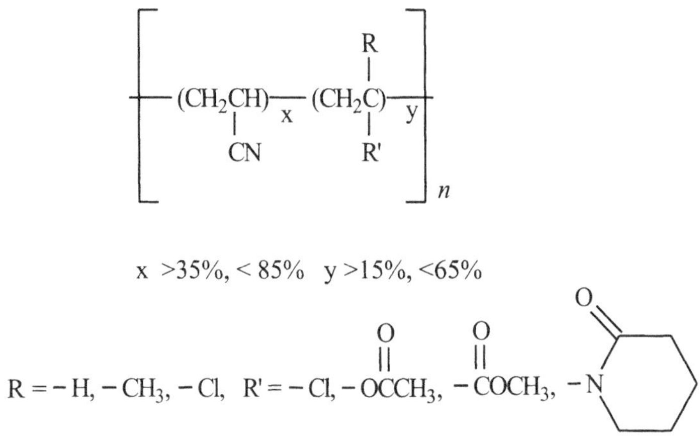

Figure 3.8. Structure of modacrylic.

Figure 3.9. Modacrylic ×1000.

3.4.3 Chemical Properties

The modacrylic fibers exhibit excellent resistance to chemical agents. They exhibit good stability to light and biological agents. Modacrylic fibers melt at 190°–210°C with slight decomposition.

3.4.4 Modacrylic End-Use Properties

Modacrylic has the following properties:

1. Staple is common due to its wool-like appearance.
2. Modacrylics possess wool-like aesthetics and a generally bright luster.
3. Crimping and spinning produces a bulky and resilient product like polyester.
4. High bulking does not relate to performance.
5. Yarn weight needs to be approximately the same as wool to perform like wool.
6. Modacrylic fiber exhibits fair pilling and abrasion resistance.
7. Bulk and abrasion resistance are similar to wool.
8. Modacrylics are resistant to attack by household chemicals and have excellent sunlight resistance.
9. The modacrylic fibers are more difficult to dye with acid and/or basic dyes than acrylics.
10. Modacrylic fibers have better affinity for disperse dyes and give dyeings of good colorfastness.
11. The fiber has poor appearance retention.
12. Static generation is low.
13. Modacrylic fibers are flame resistant and self-extinguishing on removal from a flame, with a LOI of 27.
14. Modacrylics melt at 275°–300°F.

3.5 POLYOLEFIN FIBERS

Polyolefin fibers are those fibers produced from polymers formed by the chain growth polymerization of olefins (alkenes) which contain greater than 85% polymerized ethylene, propylene, or other olefin units.

3.5.1 Polyethylene and Polypropylene

In general, linear high-density stereoregular polyethylene and polypropylene are used in textile applications, with polypropylene predominating due to its superior temperature stability. These fibers have good strength and toughness, have good abrasion resistance, and are inexpensive. The fibers are difficult to dye and have relatively low melting points, but they are effectively used in a wide variety of textile applications.

3.5.2 Structural Properties

Linear polyethylene and polypropylene are polymerized from their corresponding monomers to a DP of 1000–3000 using complex metal (Ziegler-Natta) catalysts at 40°–100°C and at moderate pressures. Using this initiation technique, branching of the polyolefin due to free radical chain transfer is avoided, and a linear unbranched structure is formed. With polypropylene, catalyst systems are selected that will lead to regular isotactic placement (>90%) of the optically active methyl-substituted carbon on the backbone to give a structure capable of high crystallinity (Fig. 3.10).

The resultant polymers are dried, compounded with appropriate additives, melt spun into fibers, and drawn to orient. The highly linear chains of these polyolefins can closely pack and associate with adjacent chains through van der Waals interactions and possess crystallinity in the 45%–60% range. The surface of these fibers is usually smooth, and the fiber cross section is round (Fig. 3.11). Also, polyolefin films can be split (fibrillated) using knife edges to form flat ribbonlike fibers with a rectangular cross section.

3.5.3 Physical Properties

Polypropylene and polyethylene are strong fibers with good elongation and recovery properties. The tenacity of fibers varies from 3.5 to 8 g/d (31–81 g/tex), with an elongation at break of ±0%–30%. The fibers recover well from stress, with 95% recovery at 10% elongation. The fibers are moderately stiff and have moderate resiliency on bending. Moisture does not affect these properties, since polyolefins are hydrophobic and have a moisture regain of 0%. The polyethylene fibers have specific gravities of 0.95–0.96 and polypropylene specific gravities of 0.90–0.91. As a consequence, these fibers float on water and are the lightest of the major fibers in commerce.

The fibers are unaffected by solvents at room temperature and are swollen by aromatic and chlorinated hydrocarbons only at elevated temperatures. They exhibit excellent heat and electric insulation characteristics and are extensively used in these applications. The fibers are heat sensitive. Polyethylene softens at 130°C and melts at 150°C, while polypropylene softens at about 150°C and melts at about 170°C.

Polyethylene

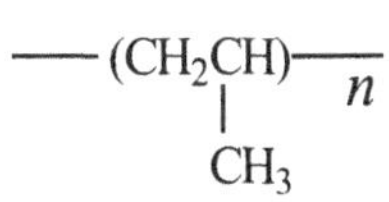

Polypropylene

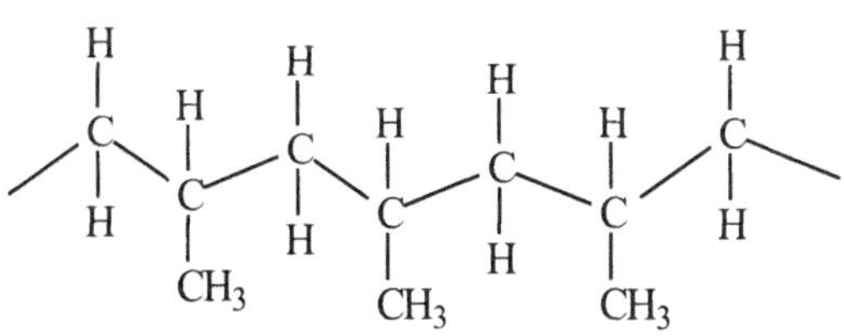

Isotactic Polypropylene

Figure 3.10. Structures of polyolefins.

Figure 3.11. Polypropylene ×1000.

3.5.4 Chemical Properties

The polyolefins are extremely inert and resistant to chemical attack. They are unaffected by chemical and biological agents under normal conditions. They are sensitive to oxidative attack in the presence of sunlight due to formation of chromophoric keto groups along the hydrocarbon chain. These groups act as photosensitizers for further decomposition. The fiber only slowly undergoes oxidative decomposition at its melting point.

3.5.5 Polyolefin End-Use Properties

Polypropylene has the following properties:

1. Strength is similar to nylon.
2. Tensile strength is very good along with abrasion resistance and durability.
3. Static build up is low.
4. Poor resilience and texture retention are common.

5. Stain resistance is excellent to waterborne stains, but less resistant to oil-borne stains, and it is not reactive to many chemicals.
6. Fibers are chemically inert and resistant to oxidative attack, except that induced by sunlight.
7. Stabilizing chemicals must be incorporated into the fibers to lower their susceptibility to such light-induced attack.
8. Polyolefins do not retain moisture.
9. The fiber is produced in continuous filament or staple.
10. Color range is limited.
11. Softening temperature is 285°–330°F.
12. The fibers are flammable, burn with a black sooty flame, and tend to melt and draw away from the flame.
13. The polyolefins have high affinity for oil-borne stains.

A summary of the properties of the man-made fibers is found in Table 3.1.

3.6 WOOL

Wool is a natural highly crimped protein hair fiber derived from sheep. The fineness and the structure and properties of the wool will depend on the variety of sheep from which it was derived. Major varieties of wool come from Merino, Lincoln, Leicester, Sussex, Cheviot, and other breeds of sheep. Worsted wool fabrics are made from highly twisted yarns of long and finer wool fibers, whereas woolen fabrics are made from less twisted yarns of coarser wool fibers.

3.6.1 Structural Properties

Wool fibers are extremely complex, highly cross-linked keratin proteins made up of over seventeen different amino acids. The amino acid content and sequence in wool varies with variety of wool. The average amino acid content for the major varieties of wool are given in Table 3.2.

Table 3.1. Physical Properties of Fibers

Fiber	Breaking Tenacity Grams per Denier		Specific Gravity g/cm_3	Moisture Regain (%)	Melt Temperature (°F)
	Dry	Wet			
Nylon 6,6 Filament	3.0–9.5	2.6–8.0	1.14	4.0–4.5	500
Nylon 6,6 Staple	3.5–7.2	3.2–6.5	1.14	4.0–4.5	500
Nylon 6 Filament	6.0–9.5	5.0–8.0	1.14	4.5	414–428
Nylon 6 Staple	2.5	2.0	1.14	4.5	414–428
Olefin Filament	4.8–7.0	4.8–7.0	0.91	0	325–335
Olefin Staple	4.8–7.0	4.8–7.0	0.91	0	325–335
Polyester Filament	4.0–9.5	4.0–9.4	1.22–1.38	0.4–0.8	480–550
Polyester Staple	2.5–6.5	2.5–6.4	1.22–1.38	0.4–0.8	480–550
Acrylic Filament	2.0–3.5	1.8–3.3	1.14–1.19	1.3–2.5	450–500
Acrylic Staple	2.0–3.5	1.8–3.3	1.14–1.19	1.3–2.5	450–500

Table 3.2. Amino Acid Contents in Wool Keratins

Amino Acid	Content in Keratin (g/100 g wool)
Glycine	5–7
Alanine	3–5
Valine	5–6
Leucine	7–9
Isoleucine	3–5
Proline	5–9
Phenylalanine	3–5
Tyrosine	4–7
Tryptophan	1–3
Serine	7–10
Threonine	6–7
Cystine	10–15
Methionine	0–1
Arginine	8–11
Histidine	2–4
Lysine	0–2
Aspartic acid	6–8
Glutamic acid	12–17

The wool protein chains are joined periodically through the disulfide cross-linked cystine, a diamino acid that is contained within two adjacent chains. About 40% of the protein chains spiral upon themselves and internally hydrogen bond to form an α-helix. Near the periodic cystine cross-links or at points where proline and other amino acids with bulky groups occur along the chain, the close packing of chains is not possible and a less regular non-helical structure is observed. The cross-linked protein structure packs and associates to form fibrils, which in turn make up the spindle-shaped cortical cells, which constitute the cortex or interior of the fiber. The cortex is made up of highly and less cross-linked "ortho" and "para" cortex positions. The cortex is surrounded by an outer sheath of scalelike layers or cuticle, which accounts for the scaled appearance running along the surface of the fiber (Figs. 3.12 and 3.13).

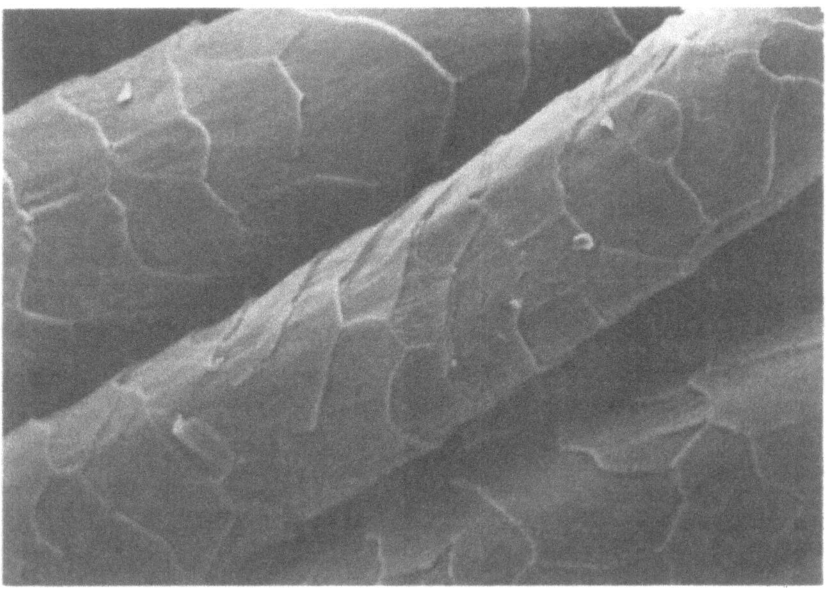

Figure 3.12. Wool (magnification × 900).

3.6.2 Physical Properties

Wool fibers possess low to moderate strength with tenacities of 1–2 g/d (9–18 g/tex) dry and 0.8–1.8 g/d (7–16 g/tex) wet. Elongations at break vary from 25% to 40% dry and 25% to 60% wet. At 2% extension, wool shows 99% recovery, and even at 20% extension, a recovery as high as 65% is observed. Wool fibers have excellent resiliency and recover readily from deformation except under high humidity. The stiffness of wool varies according to the source and the diameter of the individual fibers. The moisture regain of wool is very high and varies between 13% and 18% under standard conditions. At 100% RH, the regain approaches 40%. Wool fibers have specific gravities of 1.28–1.32. Wool is insoluble in all solvents except those capable of breaking the disulfide cross links, but it does tend to swell in polar solvents. Wool is little affected by heat up to 150°C and is a good heat insulator due to its low heat conductivity and bulkiness, which permits air entrapment in wool textile structures. At moderate humidity, wool does not build up significant static charge.

3.6.3 Chemical Properties

Wool is resistant to attack by acids, but is extremely vulnerable to attack by weak bases even at low dilutions. Wool is irreversibly damaged and colored by dilute oxidizing bleaches such as hypochloride. Reducing

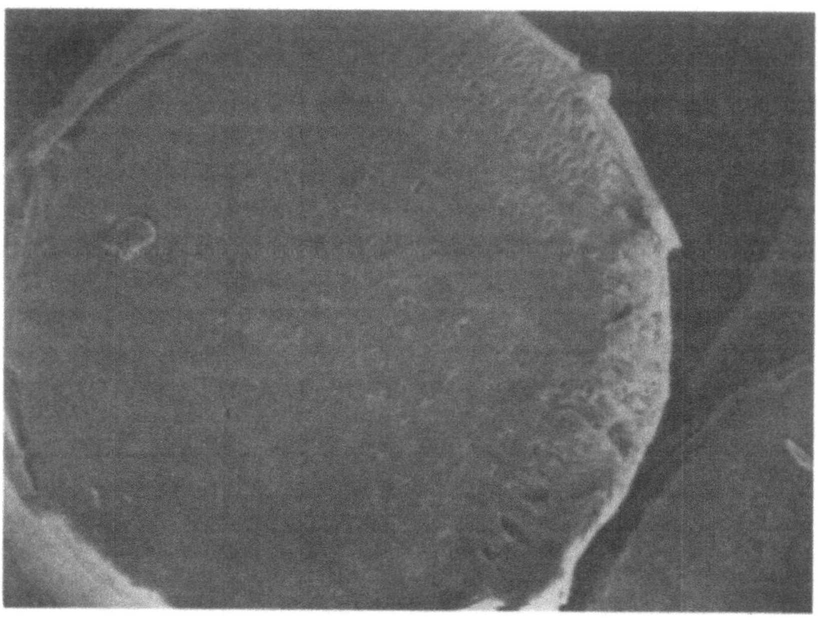

Figure 3.13. Wool cross section (magnification × 2400).

agents cause reductive scission of disulfide bonds within the wool, eventually causing the wool to dissolve. Under controlled conditions, reducing agents can be used to partially reduce the wool and flat set or set permanent pleats in the wool. Unless chemically treated, wool is susceptible to attack by several species of moths able to dissolve and digest wool fibers. Wool is quite resistant to attack by other biological agents such as mildew. Wool is attacked by short wavelength (300–350 nm) ultraviolet light, causing slow degradation and yellowing. On heating, wool degrades and yellows above 150°C and chars at 300°C.

3.6.4 Wool End-Use Properties

Wool has the following properties:

1. It is protein.
2. The outer surface has serrated scales.
3. The center is composed of cells that supply elasticity and strength.
4. The coil-like structure resists crushing and bending.
5. Wool is fairly abrasion resistant (similar to acrylic) and does not tend to form pills due to its low strength.
6. It can be stretched 25%–35% without breaking.
7. Approximately twice the amount by weight is needed to achieve the same wear performance as nylon.
8. Wool dyes readily, and the dyed wools exhibit good colorfastness.
9. There is high static build up at low humidity.
10. Soil hiding and repelling is higher than synthetic fibers.
11. Wool is a fiber of high to moderate luster.
12. Wool is resistant to attack by acids, but is extremely vulnerable to attack by weak bases even at low dilutions. Chemicals and bleach can damage.

13. Wool is attacked by alkalis and chlorine bleaches and is progressively yellowed by the short ultraviolet wavelengths in sunlight.
14. Wool is self extinguishing and has an LOI of 25.
15. Scorching occurs at 400°F and charring occurs at 570°F.
16. If a cigarette is dropped on wool, it will char easily. Once the char is removed, there is little to no trace of damage to the fiber.

Part 2

Carpet Making

4

Yarn Formation

Yarn formation methods were originally developed for spinning of natural fibers including cotton, linen, wool, and silk. Since the overall physical characteristics of the fibers and processing factors needed differed from fiber to fiber, separate processing systems were developed. As synthetic fibers were introduced, synthetic spinning systems for texturized and un-texturized cut staple were developed as modifications of existing staple systems, whereas spinning systems for texturized and un-texturized filament were developed separately. Staple yarn formation involves multiple steps and can include the following:

1. Fiber cleaning and opening (as needed for natural fibers)
2. Fiber blending (to assure uniform mixing in natural fibers or in fiber blends)
3. Carding (to align fibers and to remove short fibers)
4. Combing (if highly aligned fibers are desired)
5. Drawing and spinning (to reduce the denier of the yarn, to provide twist, and to give cohesion to the yarn)
6. Doubling or plying and twisting of the yarns (as needed to provide greater uniformity)

In recent years, a number of staple spinning processes have been developed that reduce or shorten the number of steps necessary for formation of yarns suitable for textile substrate formation. Yarn preparation from fiber filaments is much less complex, and often no, or only limited, twist is imparted prior to use in the textile substrate. The steps involved in yarn formation are outlined in Fig. 4.1.

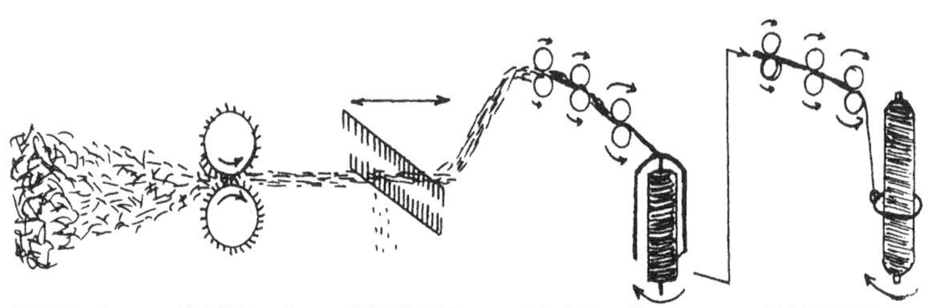

Figure 4.1. The steps in yarn formation.

4.1 WOOLEN AND WORSTED SYSTEMS

Wool is spun into yarns by either the woolen or worsted system. The woolen spinning system is less complex than the worsted system and utilizes shorter wools of a wider range of lengths and diameters to give a low twist, bulky yarn. The worsted system produces highly twisted, fine yarns utilizing fine fibers of a narrow distribution of length and size. Carpet made from woolen yarns tend to be bulky and contain more entrapped air; whereas carpet from worsted yarns tend to be tightly woven and fine with a hard flat surface.

On arrival at the mill as bales, raw wool contains large amounts of grease (lanolin), *swint* (salts from the body of the sheep), dirt, vegetable material, and other impurities. The wool must be washed in successive baths of detergent solution to remove these impurities. The process is called *scouring*, and the weight of the raw wool can be reduced by as much as 50% by the scouring process. Wool grease is effectively recovered from the scouring liquor as a commercial product. Vegetable matter remaining in the wool can be removed by passing the wool through concentrated sulfuric acid to chemically destroy the cellulosic matter, a process called *carborization*. After washing and drying, the cleaned wool is blended and carded as described previously to form a *sliver*. At this point, the sliver can be drawn and slightly twisted to form a *roving* which can then be spun into a woolen yarn. The sliver must undergo additional straightening, orientation, and removal of short fibers to be used in the worsted system. This process involves several successive steps including *gilling* (a form of pin orientation) and combing to give wool top. The wool top is drawn and slightly twisted in several stages to form a roving which is finally ring spun into a highly twisted, worsted yarn.

4.2 STAPLE SYSTEMS

Cut staple man-made fibers arrive in boxes at the mill and are ready for carding and processing into yarn. When two or more different staple fibers are mixed, extensive blending before carding is critical with repeated doubling of the sliver to assure intimate blending prior to roving and spinning. Spinning methods that take fewer steps have been developed for staple spinning and are discussed under other yarn forming methods.

4.3 FILAMENT SYSTEMS

Filament spinning systems are much less complex because the fibers are continuous and do not need to be highly twisted to give a cohesive, strong yarn. Filament yarn spinning usually involves man-made fibers and only the portion of the ring spinning system that involves twisting and winding onto spindles is used.

5

Primary and Secondary Backing Construction

Weaving and nonwoven formation are used to form primary and secondary carpet backings. The carpet yarns are tufted through the primary backing and the secondary backing to hold the tufts in place. The main technologies are outlined below.

5.1 WOVEN

Approximately 90% of all primary backings used in manufacturing tufted carpet are woven polypropylene. The reasons for its use are consistency, uniformity, low cost, ability to resist staining, enabling fine gauge tufting, and low shrinkage.

Woven primaries can be manufactured in various densities. Loosely woven primaries are made for the widest gauge tufting machine, such as 3/16 and 5/32. Denser woven primaries are used in 1/8 and 1/12 gauges.

The most widely used woven tufting primaries are made from slit film polypropylene, which are woven in the length and width of material. The largest suppliers of tufting primaries are BP Amoco, Wayn-Tex, and Synthetic Industries, and are supplied under such names as PolyBac®.

PolyBac typically has *pick counts* or the number of woven ends ranging from 11 to 22. These different pick counts allow for the use of different tufting gauge machines.

FLW is another type of woven slit film polypropylene tufting primary. A pre-dyed polypropylene fiber is added to the cap of the primary to add bulk. A nylon substitute fiber which is dyeable to match the face fiber can be used as an FLW also.

5.1.1 Woven Primary or Secondary Backing Manufacture

Yarns often must undergo additional processing before they are ready for use in the forming of a carpet. The yarns may need to be rewound, shaped, and sized onto appropriate packages; reinforced by application of sizing; lubricated by application of spinning oil; and/or drawn-in and tied into the machine used in textile formation.

Winding. Winding processes involve movement of yarn from one package to another and often conversion of the overall size, shape, and tightness of the packages. These processes also serve other important functions. Winding allows clearing of the yarn to eliminate thin spots, thick spots, knots, and other imperfections, and makes it possible to regulate tension within the package, combine or segment yarn packages, and prepare packages for dyeing prior to substrate formation. In shuttle weaving, small packages referred to as *quills* or *pirns* that fit within the shuttle must be prepared. The yarn is wound onto the pirn sequentially in such a way to assure steady and even release of yarn from the pirn during the weaving process.

Warping and Slashing. A specialized type of package formation is involved in preparing warp beams for weaving or warp knitting. A high degree of tension is placed on the warp during these processes, therefore, the yarns must be lubricated to minimize friction between yarn and machine parts, and adhesive must be applied to the yarn to strengthen and reduce the hairiness of the yarn. Warping involves winding yarns from several thousand packages placed on creels onto a flanged beam passing through a *reed* (a comblike device). The reed maintains the yarns parallel to one another as they are wrapped onto the beam under as even a tension as possible. Warping of small sections of warp (*tape warping*) is also often carried out, and the tape warps are later placed parallel to one another to provide a full width warp for use in the loom or warp knitting machine.

Staple yarns and some filament yarns must undergo *slashing*. Slashing involves simultaneous application of sizing and lubricant to the warp from one bath called a *size box*, followed by drying to remove water or solvent, breaking the slashed warp yarns away from one another using least rods, and rewinding of the warp. Sizes and lubricants used on warps will vary with fiber type. Sizes used include starches and gums, cellulose derivatives such as carboxymethyl cellulose, proteins, polyvinyl alcohol, polyvinyl acetate, and acrylic copolymers, while the lubricants used include mineral and vegetable oils and waxes, as well as derivatives of these materials.

Drawing-In and Tying-In. After warp beams are prepared, the warp yarns must be drawn through certain elements in the loom or warp knitting machine before the substrate can be produced. This process, in the past, was carried out by hand using a special hooked wire to draw each yarn through the elements of the loom or wrap knitter, followed by hand-knotting of the yarn to the corresponding yarn on the take-up warp beam. Machines are now primarily used to perform the function of drawing-in and tying-in at a high rate. Similar drawing-in of yarns for fill knitting and tufting is also necessary, but the process is not as complex as drawing-in a warp beam. In nonwoven formation, a sliver, a random, or plied-fiber web is used.

5.1.2 Textile Substrate Formation

Textile substrates are formed from yarns or fiber webs by several techniques including weaving, knitting, tufting, and nonwoven formation. In addition, composites of textile substrates are formed by methods such as adhesive bonding, formation of back coatings on substrates, and flocking. *Weaving* involves interlacing two sets of yarns, usually at right angles to one another, using a loom. The warp yarns are fed into the loom and filling (weft) yarns inserted into the warp using a shuttle or an alternative insertion technique. In tufting, yarns threaded through needles are punched through a tufting primary and the loops thus formed are held in place as the needles are withdrawn from the backing, followed by formation of the next tuft in the same manner (Ch. 6, "Carpet Construction").

In *nonwoven formation*, a fiber web or yarns are entangled or bonded to adjacent fibers through use of mechanical or chemical bonding techniques to make a continuous interconnected web. *Composites* of textile substrates are formed by bonding two substrates together using an adhesive to form a bonded substrate or backed substrate, or by application of cut fibers to an adhesive-coated substrate to form a flocked substrate.

5.1.3 Weaving

Weaving has been traditionally conducted on looms using a shuttle carrying a package (pirn) containing fill (weft) yarn which inserts the fill yarn into the warp that has been drawn-in and tied-in to the loom. In recent years, many shuttle looms have been replaced with shuttleless systems, particularly for simple constructions. The shuttle loom continues to be the most versatile weaving machine capable of weaving the widest range of yarns into primaries. The basic components of a loom are presented in Fig. 5.1.

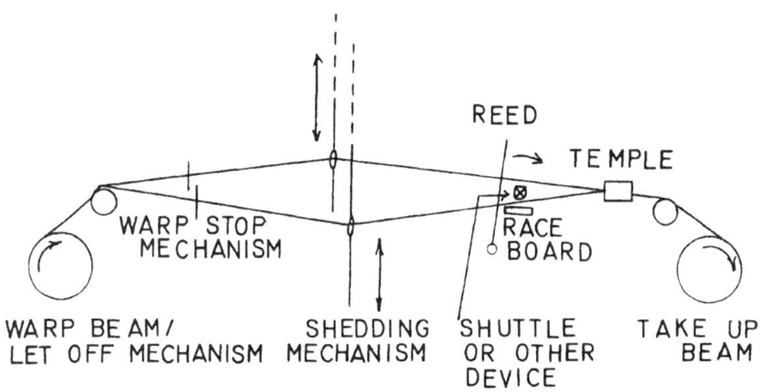

Figure 5.1. Basic components of a loom.

The loom functions in the following manner. The warp beam is connected to a let-off mechanism that meters the warp yarns off the beam as fill insertion proceeds. Each yarn in the warp passes through metal warp stop mechanisms that can detect broken warp yarns, through the eyes of heddles that are contained in the various harnesses used to lift the warp yarns, through the reed used to beat-up the filling yarn, and finally onto the take-up beam. Each warp yarn passes only through the eyes of heddles within harnesses that will be used to raise that particular warp yarn. With the addition of harnesses, the complexity of the weave can be increased. The pattern of the warp is determined by the harnesses through which each of the warp yarns is passed. The actual raising and lowering of the harnesses within the loom is referred to as *shedding*, and the space between the separated warp yarns is called the *shed.* The harnesses are raised and lowered by use of cams or a *dobby attachment,* or can also be raised and lowered individually by use of a *Jacquard mechanism.*

The fill yarn is inserted at right angles to and through the shed by use of a shuttle or an alternative insertion mechanism by a process called *picking*. Below the insertion mechanism is the race board that helps provide support for the shed and the fill insertion area. The reed is placed between the open shed and the fill insertion area of the shed. After the fill yarn has been inserted, the reed is used to push the fill yarn tightly into place in the material by a process called *beat-up*. The temple provides uniform tension on the formed web to prevent loosening of the inserted fill yarns, and a take-up mechanism keeps proper tension on the woven substrate as it is formed and taken off the loom. Therefore, the basic repeating sequence of actions in the loom is shedding to open the warp, picking to insert the fill yarn between the separated warp, and beat-up to push the fill tightly into place. The sequence must be carefully timed and synchronized to assure proper insertion. The ratio of fill insertion in a shuttle loom varies from 75 to 300 picks per minute, while the shuttleless insertion systems operate at 150 to 600 picks per minute. Fabrics woven from staple and filament yarns are found in Figs. 5.2 and 5.3, respectively.

Figure 5.2. Woven staple fabric ×50.

Figure 5.3. Woven filament fabric ×45.

5.1.4 Shedding Mechanisms

The cam system is the most limiting in the complexity of weaves possible since only 8 to 10 harnesses can be effectively raised and lowered by this method. The cams are generally positioned below the harnesses and raise and lower the harnesses by use of mechanical tappets. For more complex weaves, the dobby mechanism or the Jacquard system must be used.

The dobby mechanism is actually mounted on the side of the loom and is capable of raising as many as 20 to 28 harnesses. The dobby mechanism uses a slotted drum and continuous pattern chain containing patterns of pegs on each bar of the pattern chain to lift selected harnesses during shedding. As a bar of the pattern chain is presented to the slotted drum, the pegs present on that bar enter and block the corresponding slots in the drum. In turn, this prevents hooks attached to the harnesses from entering the slots, engaging, and raising that harness. Where the slots are not occupied by pegs, the hooks attached to individual harnesses enter the slots, engage, and the harnesses are raised. When the next bar on the pattern chain is presented to the slotted drum, a new series of harnesses are raised depending on the pattern of pegs on that bar.

In the Jacquard system, each warp yarn is attached by an individual heddle to a draw string mounted above the loom, and it may be raised and lowered independently. The draw strings from the warp yarns pass parallel to each other through separate holes in a perforated board. A series of horizontal needles mounted above the perforated board are pressed against a continuous series of cards having a pattern of perforations. Each cord is in turn attached to a rod that passes through the eye of one of these needles. The upper ends of these rods are hooked and then rest over a series of bars called *griffes*. Where the card has a perforation, the needle passes through the perforation, the hook for that warp remains on the bar, and as the bar is raised, the warp yarn is lifted. Where no perforation is present, the hook is disengaged and the warp yarn is not lifted. By this method extremely complex and intricate patterns can be developed. Due to the complexity of the mechanism, the weaving rate is much slower with Jacquard than with dobby or cam mechanisms.

5.1.5 Fill Insertion

Until recent years, fill (weft) insertion was carried out by the traverse of a shuttle, containing a package of fill (pirn), within the shed, back and forth across the width of the warp. As a result, materials produced by shuttle weaving have a *selvedge* (edge) in which the fill turns in a U at the edge of the substrate to return as the next row of fill in the fabric. Different filling yarns can be inserted by use of multiple shuttles in a magazine arrangement. Because movement of the shuttle back and forth across the substrate is necessary in a shuttle loom, a mechanism for projection (*picking*) and checking of the shuttle at both sides of the loom is required. The picking stick strikes the shuttle to provide the force necessary to accelerate the shuttle to sufficient velocity to rapidly travel across the width of the loom (picking). As the shuttle nears the other side of the loom, damping mechanisms slow and ultimately stop the shuttle (*checking*), so that it is ready for rapid return across the loom width. The picking and checking action of a conventional loom requires large amounts of energy, and a high degree of vibration is inherent in shuttle systems. In order to minimize energy consumption and machine vibration and to increase the rate of fill insertion, a number of shuttleless systems of fill insertion have been developed. Although these systems are not as versatile as shuttle fill insertion, the improved efficiency and reduced noise levels of these insertion methods make them quite suitable for less complex weaves. Ultimately, shuttleless systems are expected to be used in the production of 80% of all woven tufting primaries.

Shuttleless Systems. The shuttleless systems can be divided into two major categories: mechanical or solid systems and fluid systems. Solid systems include the use of grippers or rapiers (single and double), while fluid systems use an air jet or water jet. The major fill insert methods are presented in Fig. 5.4. The shuttleless systems all insert individual pre-measured lengths of fill yarn. Therefore, a fill package does not have to be carried across the shed, thereby greatly reducing the energy required for fill insertion. The pre-measured fill is generally only introduced from one side of the loom, and a traditional stable selvedge with the fill yarn turning back on itself is not produced. The edge of the textile web produced by the shuttleless systems is normally fringed. The fringed selvedge can be reinforced by use of a higher density of warp yarns at the selvedge or by attachments producing tucking-in of the selvedge or a leno selvedge. In shuttleless looms, the fill yarn is usually pre-measured and cut, and held in place by a vacuum tube or on a storage drum prior to insertion.

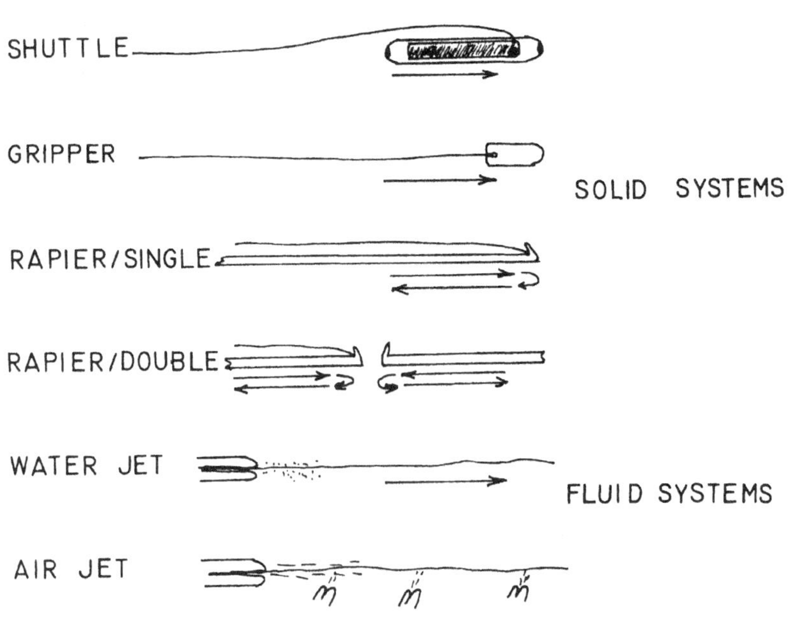

Figure 5.4. Fill insertion methods in weaving.

Gripper System. The gripper system is most closely related to the shuttle system. The gripper is fired as a projectile across the width of the loom carrying a single length of fill. The gripper is much smaller and lighter than a shuttle since it does not need to carry a fill package. Multiple grippers are used, and a gripper after insertion of fill yarn is returned by a conveyor system back across the loom. A multiphase gripper system is also used in a series; each gripper carrying fill yarn is conveyed across that loom in sequence by use of a magnetic or mechanical drive mechanism. This method must use phased shedding and beat-up motions to permit simultaneous movement of several grippers across the face of the loom. Although the velocity of the multiple grippers across the loom is much slower than that of a single gripper, the composite rate of fill insertion is much faster than more conventional looms.

Rapier Systems. In the rapier systems, the fill yarn is carried across the warp by a single arm or two mechanical ones. The rapiers must be removed from the shed prior to beat-up. In the single rapier system, the end of the arm contains a clip to hold the fill yarn that releases the yarn after the fill yarn is completely inserted. In the double rapier system, one arm equipped with a clip (giver) conveys the yarn to the middle of the shed, and the taker on the other arm simultaneously is inserted from the other side and takes the yarn across the rest of the shed. Since rigid rapiers effectively double the width of the loom, flexible rapiers that uncoil upon fill insertion have been developed that reduce the loom width.

Fluid Fill Insertion Systems. Fluid fill insertion systems do not use a device to carry the fill yarn across the loom. They operate by impinging sonic velocity water or air jets onto the end of the yarn, which accelerates and carries the yarn across the loom. Since water is more cohesive than air and the energy conferred to the water is not as readily dissipated as the energy conferred to air, the water jet is capable of conveying the fill yarn greater distances than an air jet. The major disadvantage of water jet fill insertion methods is related to the hydrophilic character of the water and its ability to dissolve many sizes and to wet out hydrophilic fibers. Therefore, water jets can only be effectively used on hydrophobic fibers such as polyester that are unsized filaments or that contain sizing unaffected by water. To enhance the projection distance of air jet systems, guides are mounted across the loom that are inserted through the warp during fill insertion to provide a turbulence-free path across the loom. Also, booster jets are often mounted periodically across the loom and fired sequentially as the fill yarn is inserted to assist in carrying the yarn across the width of the loom.

5.2 NONWOVEN

Most nonwoven primary backings are made of polyester, polyester/ polyamide blends, or polypropylene. These primaries are mainly used in carpets that require technical performance. Automobile carpet, which requires mold ability, and carpet tiles, which require dimensional stability, are two examples of specialty uses.

There are three major nonwoven tufting primaries used in carpets. These are Typar®, Lutradur®, and Colback®. Typar is made of polypropylene. Colback is produced by thermally bonding spun laid bi-component filaments. These filaments have a polyester core and a polyamide or nylon surface. Thermal bonding takes place at the intersection of the filaments by heat and pressure. Lutradur is a 100% polyester spunbonded nonwoven.

The most expensive nonwoven is made from polyester. Polyester has a higher temperature resistance than polypropylene, which is important in manufacturing processes that require thermal stability, such as carpet tiles and automotive carpeting.

Nonwoven polypropylene is not as flexible as woven polypropylene, therefore, it has fewer tendencies to bow and skew. It is used to make a lightweight level loop carpet or in printed or patterned carpets.

5.2.1 Nonwoven Primary or Secondary Backing Formation

Nonwoven textile substrates can be formed through entanglement and/or bonding of fibers in the form of webs or yarns by various chemical and mechanical means. Such methods used to form nonwovens minimize greatly the number of steps required to go from the fiber to the finished substrate and greatly reduce the cost of production of the substrate. The strength, flexibility, and utility of the resulting nonwoven substrates are generally less than that of a similar woven or knitted substrate. Therefore, nonwoven materials have only been used in apparel for interfacing or as felts, but have found extensive use in medical, industrial, and home furnishing applications. Examples of a nonwoven felt and an adhesive-bonded, nonwoven are found in Figs. 5.5 and 5.6.

The fiber web or, in some cases, films or yarns used in nonwoven formation are prepared by conventional processes. Web formation is carried out by carding, *garnetting* (a process to convert textile waste to fiber, then carding), air laying, or wet laying of staple fibers. The web can also be

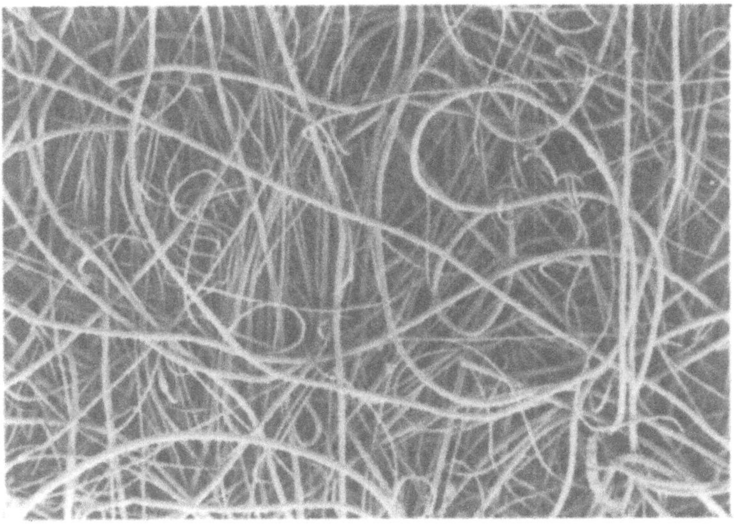

Figure 5.5. Non-woven felt ×35.

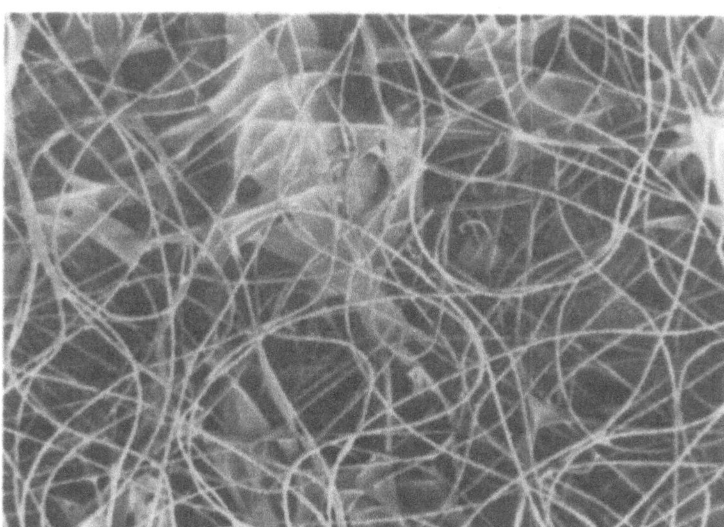

Figure 5.6. Non-woven, adhesive-bonded material ×35.

formed by direct laying of extruded filaments onto a moving belt or through extrusion of a film followed by slitting, embossing, and stretching. The webs are laid in a random or oriented fashion with oriented webs being parallel, plated, or crosslaid in one or more plies. The webs or in selected cases, yarns are entangled or are bonded together by four basic processes: mechanical entanglement, stitching, self-bonding, and adhesive bonding.

5.2.2 Mechanical Bonding or Entanglement of Nonwovens

A number of processes exist for mechanically entangling or bonding a web to form a nonwoven substrate. One method that is mechanical in nature does not involve a web, but rather splitting of a polyolefin film. In this process, a film is extruded followed by either a discontinuous pattern cut by knife slitting or by embossing the film followed by biaxial orientation to form a network structure.

Mechanical Bonding. Methods that involve mechanical bonding or entangling of fiber web include wet laying, felting, needling, and air or water jet techniques. Wet laying essentially uses the process used for making paper. A slurry of short staple fibers in water or solvent is continuously laid onto a moving wire screen. The remaining water or solvent in the fiber mass that does not fall through the screen is removed by squeeze rolls, and subsequent drying forms a randomly oriented nonwoven that can be increased by adding chemicals or binders to the slurry during the process.

Felting. Felting involves the entanglement of fibers in such a way that the fibers ratchet on one another and make the web denser. This is accomplished by using the natural felting action of scaled keratin fibers such as wool in the presence of moisture and mechanical action, or by heat-induced shrinkage of fibers like polyester accompanied by mechanical action.

Needle punching. Needle punching of a web involves passing the web between bands of needles which continuously punch through and withdraw from the web. The needles have reverse barbs and cause reorientation of some fibers in the web and complex entanglement of the web in the vertical, as well as horizontal direction. Needle punching from both sides of the web provides a stronger substrate.

Water jets or air jets. Water jets or air jets can also be used to entangle a web. The fiber web is passed under a series of high velocity water jets that are programmed to deliver short bursts of water or air at intervals. The jets disrupt the web and form bonds through mechanical entanglement of fibers in the web. In general, mechanical bonding and entanglement methods give soft hand with low to moderate mechanical strength.

5.2.3 Stitching or Stitch Bonding

Yarns or fiber webs are bonded together by stitching the webs together by a series of needles that operate very similarly to conventional warp knitting machines. Therefore, guide bars must be used as part of the mechanism to interconnect the threads being sewed into the yarns or web being fed into the machine. Stitch-bonded fabric substrates are usually made using filament thread to form a reasonably strong, but quite flexible structure. Yarns and webs can also be crosslaid in these stitch-bonded processes. The processes operate at very high operating speeds using inexpensive yarns or webs with fabric substrate being formed at 50 to 200 meters/hr.

Stitch bonding of yarns is usually carried out on Malimo® machines, whereas web stitching is carried out on Maliwatt® or Arachne® machines. Pile stitching machines such as Malipol®, Araloop®, and Locstitch® have also been developed for formation of pile fabrics.

5.2.4 Self Bonding

Techniques have been developed to bond thermoplastic fibers to each other by use of heat and/or solvent. In some cases, low levels of binder are added to the substrate prior to bond formation to assure a higher level of bonding. *Spunbonding* is the most widely used of these techniques. In spunbonding, a web of continuous filament thermoplastic fibers, such as polyester and polyolefin, are extruded and randomly laid onto a moving conveyor belt and subjected to heat and pressure between embossed rolls to form periodic bonds between fibers in the web. The flexibility of the web formed by spunbonding is determined by both the density of the web and the number of bonding points per unit area. *Melding* involves use of a specially prepared web of sheath/core filaments that can be subsequently bonded by use of heat. The sheath/core fibers have a lower melting polymer substrate making up the sheath than contained in the core. Passing the web between embossed rolls under pressure at a temperature above the softening point of the sheath polymer causes the fibers in the web to be bonded at crossover points in the structure. Since the cores of the fibers are unaffected by the treatment, the resulting bonded web is quite strong. A third method, called *gel bonding*, involves use of a solvent applied to the fiber web. Solvent is metered onto the fiber web causing softening and gelation of the fiber web, and the web is then passed through embossed pressure rollers. The gelled surfaces of the fibers merge at crossover points and, on solvent removal by heating, form bonds between fibers. The above processes form fabric substrates of moderate stiffness.

5.2.5 Adhesive Bonding

Application of adhesive to the fiber web followed by pressure and heating to cure the binder and bond the fibers in the web to one another is an extensively used method of nonwoven formation, particularly for non-thermoplastic fibers such as cellulosics. The adhesive is applied as a solution, suspension, or emulsion to the fiber web by padding, spraying, printing, or by application as a foam. The adhesives used are usually thermosetting or thermoplastic (hot melt) adhesives. After application, the web is subjected to heat to drive off solvent (usually water) and to melt and soften a thermoplastic adhesive, or to cause reaction or curing of a thermoset adhesive. The adhesive, on curing, tends to concentrate at the fiber crossover points due to capillary action. Nonwoven formation by adhesive bonding tends to form moderate to stiff substrates with the degree of stiffness depending on the nature and density of the fiber web, the type and concentration of adhesive used, and the number of bonds per unit area.

5.3 COMPOSITE FORMATION

Flexible composite textile substrates are formed by either coating textile substrates with a continuous polymer layer or lamination of two or more textile substrates together by use of an adhesive polymer layer. The polymer coating can be applied neat, from solution, as an emulsion, or in the form of a film or thin foam. Inflexible fiber-polymer composite substrates are formed by imbedding fibers, fiber webs, fiber tows, or fabrics in a stiff polymer matrix. Inflexible composites have found extensive use in engineering and aerospace applications particularly where high performance properties are important. Owing to their inflexibility, they lose any properties characteristic of textile substrates.

In order to get a good bond between the fabric substrate and the polymer coating in laminated and coated fabrics, a good adhesive bond between the textile substrate and the coating or adhesive is essential. This is particularly important since the substrate and polymer may have very different stretch and recovery properties. Adhesive failure under stress is the most probable cause of failure in these materials. In coated fabrics, elastomeric properties are necessary to achieve a serviceable coated substrate.

In coated substrates, the polymer coating is metered and spread evenly onto the fabric surface as a controlled viscosity solution, aqueous emulsion, or hot melt polymer; followed by drying; and curing if needed. A urethane foam can be bonded to a textile substrate by carefully melting the foam surface with a flame, joining the melted foam surface to the textile substrate under pressure, followed by cooling. Laminated fabrics are bonded under similar conditions except that two fabric faces are brought together and bonded by a polymer layer.

5.4 NEW TUFTING PRIMARY BACKINGS

New tufting primary backings have been developed with a combination of woven and nonwoven materials. This combines performance attributes of both woven and nonwoven tufting primaries. Therefore, these backings are used primarily in carpet, which requires high performance.

6

Carpet Construction

The carpets in today's commercial and residential installations are a compilation of layers configured to meet the requirements of each site. Carpet designers take in to account, among other factors, the wear and tear the carpet will receive, the need for sound absorption, the exposure to environmental conditions such as sunlight, and how the layers will work together to achieve the desired goals. This chapter lays the foundation for understanding how the designer meets the specifications for a carpet by explaining the layers that go into making a tufted carpet.

6.1 DESCRIPTION OF LAYERS OF PILE CARPET

6.1.1 Primary Backings

Primary backing is a material into which tufts of yarn are inserted to create pile. Both primary and secondary backings used for carpets are made primarily from polyolefin and polyester in the form of woven and nonwoven materials.

6.1.2 Loop Pile Created by Tufting

Pile for tufted carpets is made on a tufting machine (Fig. 6.1). *Tufting* is a rapid process for the formation of continuous rows of yarn loops across the face of a textile backing substrate. The yarn loops on the tufted substrate then can be cut or sheared to form a cut pile, if desired.

Nylon yarns account for over 70% of the yarns used to form the tufted face of the substrate, with polyester, polypropylene, acrylic, modacrylic, and wool yarns being used to lesser extents. Nylon dominates the tufted carpet market due to its overall toughness and resiliency.

Figure 6.1. A tufting machine used for making residential carpet. *(Courtesy of Lyle Industries.)*

In tufting, a woven or film primary backing material is passed under a row of tufting needles each carrying a yarn. The needles penetrate the backing to form a loop that is held in place as the needle is withdrawn through the backing. Subsequent loops are made in the same manner to form continuous rows of tufted loops on the substrate.

After the loops are formed, all or some of the tufted loops may be cut by special cutting devices.

After the tufted yarn has been inserted into the backing, a layer of adhesive or coating must be applied to the backside of the tufted structure to mechanically fix the tufted yarns in place. Often a secondary backing in the form of a textile substrate is placed and fixed by adhesive over the back side of the yarns, or polyurethane foam backing is applied to the back of the tufted substrate. The basic elements of the tufting process are found in Fig. 6.2.

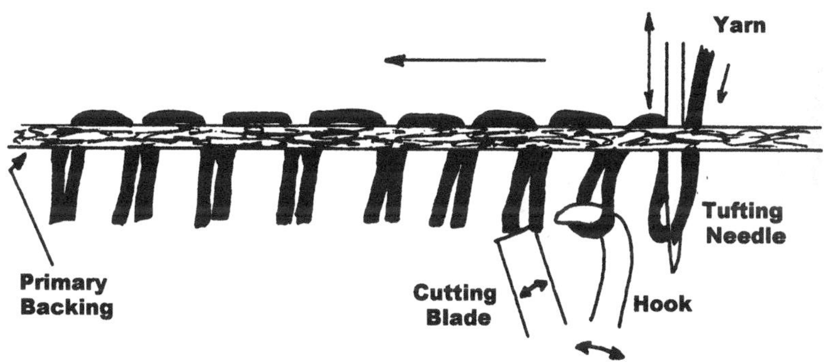

Figure 6.2. Basic elements of the tufting process.

The easiest way to visualize the tufting process is to think of a sewing machine that is large and wide, utilizing hundreds of needles. The yarn used in tufting can be thought of as thread in a sewing machine. Yarn is placed on a creel or beam behind a tufting machine. Each yarn is fed through a guide or tube to each needle. Pull rollers take the yarns out of the guides and feed them into the tufting machine. The pull rollers control the amount of yarn supplied to the machine.

There are guides within the tufting machine itself. These guides are needed to prevent yarns from becoming tangled in the machine. Yarns travel through a jerker bar (Fig. 6.3), which is a combination of guides that are attached to the tufting machine and on the top of the machine, the arm extending from the needle bar. Slack in the yarn occurs during the upward movement of the needles and is removed by the two guides.

After the jerker bar, the yarns travel through the needles that are attached to a needle bar. Each needle is evenly spaced in a row across the bar (Fig. 6.4).

In the tufting process, a primary backing is fed into the machine and yarn is punched into it. Typically spiked rollers are used to pull the primary backing over a bedplate. These spiked rollers control the stitch rate of the carpet. Needles penetrate the primary backing at the bedplate. Directly underneath the bedplate are loopers for loop pile or looper and knife combinations for cut and cut/loop carpet. Loopers grab the yarn from the needles. The looper holds the yarn for a set period of time and then releases it as the needle is removed from the primary backing. This forms the loop of the pile. The holding time of the yarn by the loopers determines the pile height.

Loop Pile. There are three different types of loop pile. They are level loop (Fig. 6.5), pattern loop, and textured loop. All level loop carpets have the same pile height from row to row. Carpet with different pile heights in a pattern is referred to as a pattern carpet. Slight variations in pile height, but not a definite pattern, produce a textured loop surface.

Cut Pile. Typical cut pile carpet (Fig. 6.5) has a level surface since all the yarn has the same pile height. Shifting a needle bar or using different colored yarns can achieve various patterns. For cut pile carpet, the loopers hold and cut the yarn in one movement. A patterning device is used to produce a cut/loop pile carpet. The device utilizes a special looper and conventional cutting knife.

Cut/Loop. The most common cut/loop is a high/low construction, where the high loop is cut and the low loop is left uncut. Leaving the cut yarn and loop yarn at the same pile height produces a level cut/loop carpet.

Tufting leads to a wide range of looped pile and cut pile carpet substrates with the number of tufts per unit area, the length of loops or pile, and the fiber type affecting the nature and performance of the carpet.

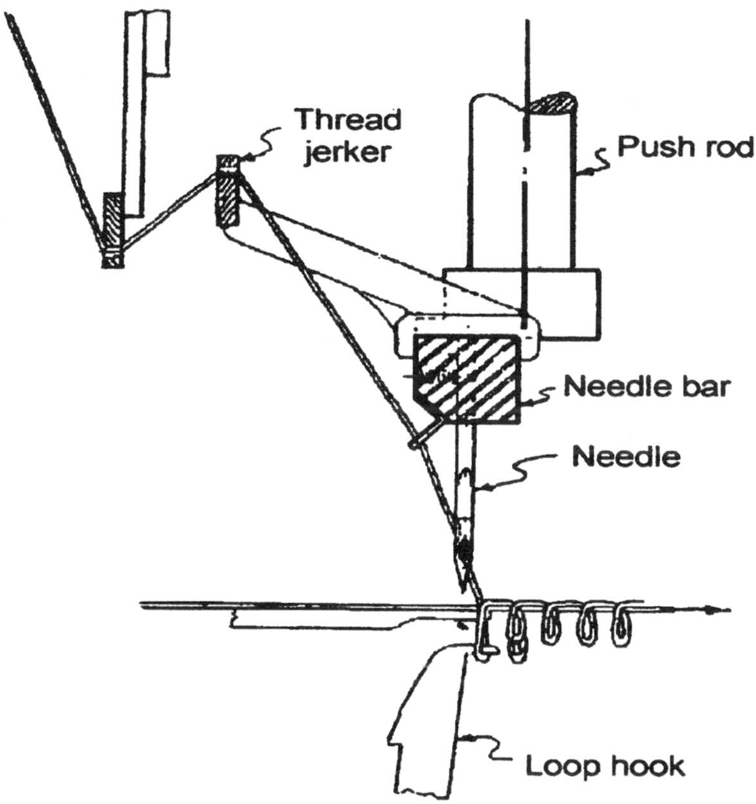

Figure 6.3. Threading of a tufting machine. *[Courtesy TUFTCO Corp. (1976).]*

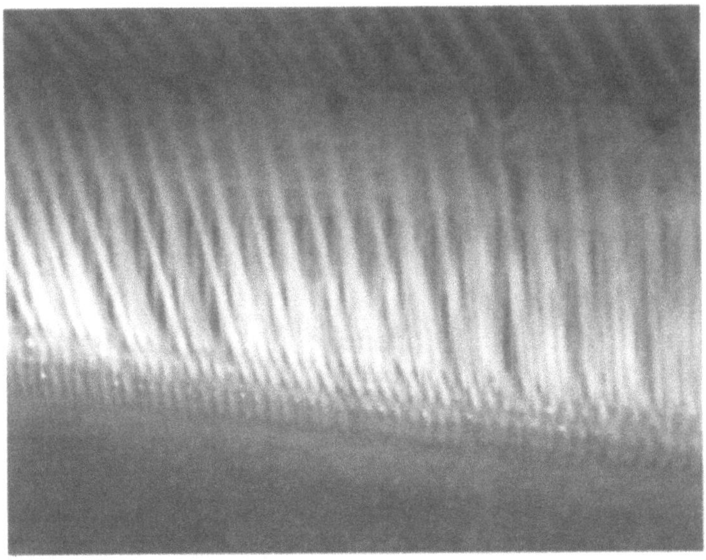

Figure 6.4. Needles on a tufting machine used to make carpet. *(Courtesy of Lyle Industries, Inc.)*

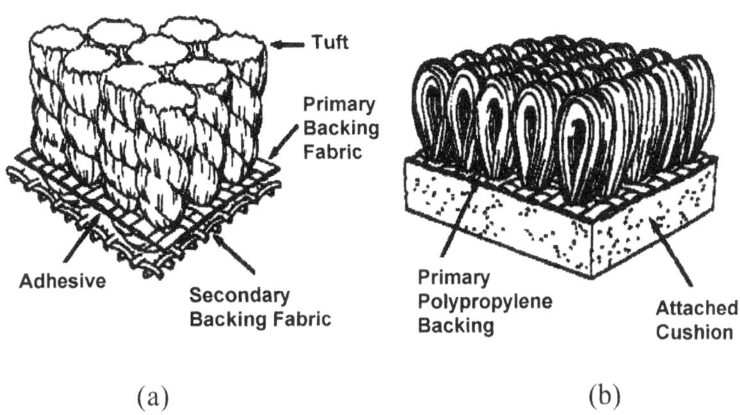

Figure 6.5. Examples of the layers of carpet and two types of pile; *(a)* cut, *(b)* loop. *(Courtesy of http://www.carpet_rug.com/pdf_word_docs_/primer_teaser/primer_section_1.pdf)*

6.1.3 Shearing

Shearing or cutting is used to remove loose fibers and clean the surface of the carpet. Many cut pile carpets made from staple fibers have a hairy or fuzzy appearance on the surface. Loose fibers also come to the surface during other carpet manufacturing processes. The choice of shearing carpet before or after the latex coating is applied is left to the carpet manufacturer. This depends upon what the carpet manufacturer believes is the best method to achieve a uniform product.

Shearing machines use heavy brushes to lift the carpet pile so that it stands erect prior to the carpet traveling through a set of spiral cutting knives. Tips of the pile are removed at controlled heights set by the shearing blade and nose bar of the shearing machine. The nose bar is an inverted V-shaped roller bar. This bar opens up the pile yarn and allows it to be cut more easily. A vacuum attachment removes the cut fibers from the shearing head. Multiple shearing heads, operating with each other, allow maximum speeds to be achieved. The best finish is obtained by reducing the amount of yarn each shearing machine is required to cut to achieve the desired finish.

Tip sheared carpet is manufactured by tufting carpet with high and low loops. During the shearing process, the high loops are sheared off to achieve a desirable appearance.

Multiple shearing heads, operating with each other, allow maximum speeds to be achieved. The best finish is obtained by reducing the amount of yarn each shearing machine is required to cut to achieve the desired finish.

Tip sheared carpet is manufactured by tufting carpet with high and low loops. During the shearing process, the high loops are sheared off to achieve a desirable appearance.

6.1.4 Secondary Backing

Any material laminated to the back of carpet is referred to as a *secondary backing* (Fig. 6.6 and see also Fig. 6.5). Secondary backings can be either woven, nonwoven, or a cushion. The application of the secondary backing improves the dimensional stability of the carpet. The reasons for applying a secondary backing are as follows:

- Hold the tufts or yarns in place

- Provide dimensional stability
- Improve resiliency
- Provide thermal insulation
- Provide sound insulation
- Create cushioning
- Improve the life of the carpet
- Reduce edge fraying
- Reduce pile shedding

Most secondary backings use a woven material made by leno weaving slit film polypropylene in the length of the construction and polypropylene spun yarn in the width. This construction is used to minimize unraveling and fraying.

6.2 FINISHING

Rolls of carpet are sewn together to form a continuous web. The web typically travels through a steamer to remove any wrinkles in the carpet and to add bulk to the yarn. A compounded latex coating is then applied to the back of the carpet. If a secondary backing is not applied, the carpet is considered to have a unitary backing.

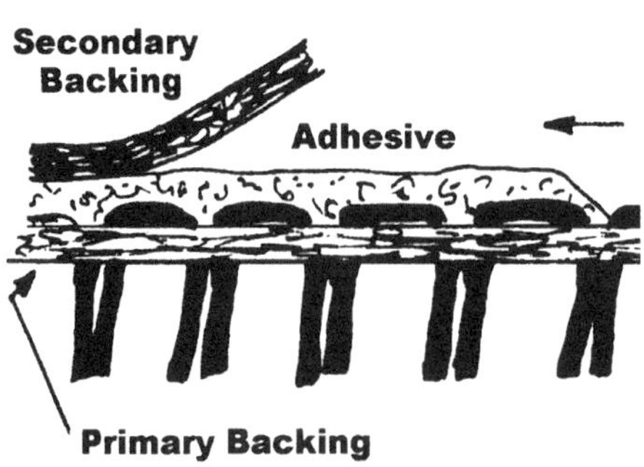

Figure 6.6. Application of a secondary backing.

In most cases, there will be a secondary backing applied to the carpet. The secondary backing has compounded latex applied to it (See Ch. 7, "Latex Coating") at the same time the latex is applied to the carpet. Before the carpet enters the oven to dry the latex, the secondary is pressed to it. Tenter pins hold the carpet on each side as it is taken through the oven. This keeps the carpet at its full width as it is processed. Once the carpet exits the oven, it is cut into rolls.

Part 3

Coatings, Raw Materials, and Their Processes

7

Latex Coatings

With the exception of unitary latex-backed carpet, all carpets require a secondary backing. To apply a secondary backing, adhesive latex coating must be used (Fig. 7.1).

Most carpet mills work with their suppliers to optimize latex compounds for cost and performance. The main goal of the carpet producer is to have the maximum performance in tuftbind, pilling or fuzzing, adhesion to the secondary (if required), dimensional stability, moisture resistance, and low flammability.

7.1 LATEX COMPOUNDS

Most carpets produced are coated with carboxylated styrene butadiene rubber (SBR) latex. Styrene butadiene rubber (SBR) is a synthetic rubber produced by copolymerization of about three parts butadiene and one part styrene. The finished elastomer is suspended in emulsion form in water in the presence of a soap or detergent.

Figure 7.1. A puddle of latex is applied to the back of a carpet by a blade that forces the latex into the back of the carpet. A second coat of latex will be applied to hold the secondary backing onto the tufted material. *(Courtesy Lyle Industries, Inc.)*

The widespread use of SBR is based on its cost effectiveness when compared to other latex types. With SBR, the sum of raw material costs, ease of compounding, processing feasibility, and quality of the finished products result in a cost performance that other latex products can not match. In addition, SBR latex is strong and can be extended with large amounts of minerals, typically calcium carbonate. Only a few specific property requirements in the carpet industry challenge SBR in performance, i.e., moisture resistance and dry cleaning.

The following are the raw materials used for latex coating:

1. Latex
2. Fillers
3. Surfactants
4. Thickeners
5. Water
6. Flame Retardants
7. Miscellaneous

The factors that contribute to the performance of the latex compound include percent solids, viscosity, pH, surface tension, particle size, and styrene-to-butadiene ratio. Filler average-particle size and particle size distribution can also affect performance. Table 7.1 shows how performance is affected by changes in the latex compound. Therefore, a compounder of latex must be careful in the selection of the mineral extender.

Table 7.1. Latex Compounds and their Typical Effects on Carpet Performance

Increasing	Tuftbind	Pilling and Fuzzing	Delamination	Flexibility
Latex Coating Weight	Improves	Improves	Improves	N/A
Yarn Bundle Wrap	Improves	N/A	Improves	N/A
Yarn Penetration	N/A	Improves	N/A	N/A
Filler Loading	Decreases	Decreases	Decreases	Decreases
Styrene Level	Improves	Improves	Improves	Decreases
Yarn Density	Decreases	Decreases	Improves	N/A
Carpet Density	Decreases	Decreases	Improves	N/A
N/A-Not available				

By increasing the level of styrene in the latex, the dried latex will increase in tensile strength, modulus, and glass transition temperature along with improving mechanical and chemical stability during compounding. An increase in resistance to solvents and oxidation will also occur.

There are other latex materials used in the carpet industry. These aqueous materials include such polymers as vinyl acetate ethylene (VAE), polyvinyldene chloride (PVDC), acrylic, and polyurethane. These polymers are designed for specific markets and information about them is scarce.

7.1.1 Filler

Fillers are used to reduce the cost of the latex, increase flame retardancy of it, control its viscosity, and so forth. Calcium carbonate and barium sulfate are mined underground or from a quarry. The minerals are crushed and pulverized to various particle sizes depending on the end-use requirements.

Aluminum trihydrate (ATH) begins as bauxite ore. During the manufacture of aluminum metal, aluminum trihydrate is produced as an intermediate.

Calcium carbonate is typically used as a filler due to its low cost. Table 7.2 provides information on common fillers used in carpet coatings.

Table 7.2. Properties of Fillers for Latex

	Calcium Carbonate	**ATH**	**Barium Sulfate**
Thermal Conductivity (Cal/g · s · °C) · 1000	5.6	0.2	5
Specific Heat	0.205	0.19	0.11(Cal/g · °C)
Young's Modulus	2.6	3	3[Kg/(cm^2 · 0.000001)]
Poisson's Ratio	0.27	0.3	0.3(*dw/dl*)
Hardness	2.5–3	2	3–3.5(Mohs)
Dielectric Constant	6.14	7	7.3($e = D/E$)
Density	2.7	2.4	4.4(g/cm^3)
Coefficient of Thermal Expansion	10	4–5	10

7.1.2 Surfactants

Sodium lauryl sulfate, ammonium lauryl sulfate, sulfosuccinamate, long chain alcohols, etc., or blends of these chemicals are used to aid in the penetration of air into the latex compound to control add-on weights or penetration into the yarn.

7.1.3 Thickeners

Polyacrylate thickeners are primarily used for viscosity control. The viscosity of the latex is used to effectively control the placement of the latex onto the carpet back.

7.1.4 Water

Water is added as needed to control the total solids requirement of the latex compound.

7.1.5 Flame Retardants

For the carpet industry, aluminum trihydrate (ATH) is the flame retardant of choice due to its low cost and performance. Aluminum trihydrate contains water molecules within its chemical structure. For more information on ATH, see Sec. 7.5.

7.1.6 Miscellaneous

Miscellaneous additives include penetrants, defoamers, dispersants, biocides, antimicrobials, antiblistering agents, and so forth. These materials are placed into the latex while it is being compounded.

7.2 EXAMPLES OF LATEX COMPOUNDS

Depending on the final performance requirements of the carpet, different levels of filler are used. Lower-filled latex will have more strength than a highly filled one.

The example in Table 7.3 is a modification of Honeywell information available to the carpet industry. For illustrative purposes, examples of latex compounds are given along with theoretical end uses based on performance requirements.

Table 7.4 lists typical applications of the latex compounds described in Table 7.3.

Table 7.3. Filler Effect on Carpet Performance

Example	Latex Type	Filler Loading (phr)*	Coating Weight (oz/sq yd dry)
1	Unitary	150	32
2	Precoat	200	20
3	Precoat	350	18
4	Precoat	450	18
5	Adhesive	350	6
*phr refers to parts per hundred resin. (*Data courtesy of Honeywell.*)			

Table 7.4. Use of Latex Compounds on Commercial and Residential Carpet*

Carpet Construction	Compounds used from Examples
Commercial Carpet	
Unitary	1
Loop	2 & 5
Graphic	3 & 5
Cut Pile	4 & 5
Residential Carpet	
Graphics	3 & 5
High Cut Pile	3 & 5
Medium Cut Pile	4 & 5
Low Cut Pile	5
Berbers	2 & 5
**Data courtesy of Honeywell*	

7.3 EFFECT OF FILLER ON LATEX-COATED CARPET

To show the effect of two significantly different calcium carbonates on mechanical properties and their economic impact on latex compounds, a test was performed. A control SBR latex was prepared with an inexpensive 200-mesh size calcium carbonate. It was compared to a more expensive latex compound made with a calcium carbonate called calcite. The particle size of the calcite was the same as the control. The latex compounds were tested for viscosity, delamination strength to a secondary backing, and overall cost. The data from the test is reported in Table 7.5.

Table 7.5 shows that cost should not be the determining factor in raw material selection. Besides the almost identical cost of the calcite to the inexpensive calcium carbonate, the higher filler loading should allow for higher processing speeds. The reasoning behind this is less drying time for the carpet with the calcite-filled latex with the higher filler loading.

7.4 EFFECT OF DENIER ON TUFTBIND

Yarn denier has a major impact on the tuftbind performance of carpet. Table 7.6 is a comparison of tuftbind using two different latex compounds applied at 30 oz/sq yd to the back of carpet. The compounds are a unitary formulation using 150-phr calcium carbonate and a standard compound using 275-phr calcium carbonate. Latex viscosity and air-to-latex ratio were kept the same.

The influence on tuftbind by yarn denier can also be expected with other polymers used to coat carpet. This is also shown in Table 7.6.

While the performance of tuftbind is affected by yarn denier, there are other variables. Some are carpet construction, latex viscosity, filler loading of the latex, air-to-latex ratio, stitches per inch of the yarn into the carpet tufting primary, and the gauge of the tufting machine.

Table 7.5. Effect of Filler on Carpet

	Control	Calcite
Cost ($/lb)	0.04	0.11

Phr filler	Viscosity (cps)	Viscosity (cps)
200	725	425
250	N/A	625
300	N/A	875

Phr filler	Delamination (lb/sq in)	Delamination (lb/sq in)
200	10.2	10.1
250	N/A	10.0
300	N/A	9.0

	Control	Calcite
Phr	200	300
Viscosity (cps)	725	875
Delamination (lb/sq in)	10.2	9.0
Compound Cost ($/lb)	0.225	0.23

Table 7.6. Effect of Yarn Denier on Carpet Performance

Avg. Yarn Denier	Tuft Gauge	Unitary Compound Tuftbind (lbs)	Standard Compound Tuftbind (lbs)
5,000	1/8	22	18
4,000	1/8	20	16
4,000	1/10	18	15
4,000	5/64	16	13
3,500	1/10	18	15

7.5 FLAME RETARDANCY

The yarn, tufting backing, latex coating, secondary backing, and styling of the carpet affect carpet flammability. Flame retardants for carpet must be chosen for their compatibility with other components in a latex compound. At the same time, the flame retardant cannot adversely affect the performance of the carpet over time.

Aluminum trihydrate is the principal flame retardant used in the carpet industry. Water is chemically bonded to ATH. This water constitutes approximately 35% by weight of ATH. When ATH is exposed to high heat, it releases water from its chemical structure. This endothermic release of water, which begins at 400°F, causes a heat sink during combustion of carpet. The release of water prevents the ignition of other combustible material. The reaction is

Eq. (7.1) $2\ Al(OH)_3 \rightarrow Al_2O_3 + 3H_2O$

When ATH is used as a flame retardant, it is used as a 25% to 50% replacement of calcium carbonate. Due to its high cost compared to calcium carbonate, the substitution is enough to meet flame retardancy requirements without adding excessive cost to the carpet.

7.6 SUMMARY

Other adhesives used to bond fibers in carpet include hot melts and reactive polyurethane. Due to their high costs, they are used in high performance areas where high tuftbind strength and moisture resistance are required. Chapters 9, 10, and 11 discuss these coating methods.

8

Polyurethane Coating

There are two basic methods of producing a polyurethane-backed carpet. The precoat can be a latex or polyurethane for either case. One backing is a mechanically frothed slow-reacting foam compound, which forms flexible foam at a selected thickness. It is cured at 175°–275°F. The other method uses a water-blown foam. Infrared heaters or convection ovens heat the reacting material to prevent heat loss and to control the time the surface is cured. Both systems allow a single pass process during manufacture.

If polyurethane is used as a precoat, the precoat typically will produce higher tuftbind, abrasion resistance, resistance to heat aging, resistance to compression, and improved sound insulation compared to styrene butadiene rubber (SBR) latex or foam compounds.

8.1 POLYURETHANE RAW MATERIALS AND BASIC CHEMISTRY

Polyurethane is formed by the reaction between an isocyanate and a polyol. For the residential and commercial carpet industry, the raw materials and chemistry are basically the same. There are three basic components used:

1. Polyether glycol mixed with a catalyst(s), silicone surfactant, and fillers (Sec. 7.1.2) to make a polyol compound.
2. A polymeric methylene di-para-phenylene isocyanate (MDI).
3. A blowing or foam-forming agent, such as water, air, or inert gas.

Since its introduction, polyurethane foam for carpet use made with a polyol with 10%–20 % ethylene oxide has proven to be a stable foam. The molecular weight of the polyol is a factor that affects the performance of the cushion; a range of 4,500–7,000, using a polyether triol, is recommended. A silicone surfactant, that can maintain stable foam, is selected based on the supplier's recommendation, as is the catalyst system. Silicone surfactants are usually polyetherpolysiloxane, and catalyst blends normally contain at least one amine component.

Methylene di-para-phenylene isocyanate is the most common isocyanate used in the production of polyurethane carpet cushion. It is safer to use than the traditional diisocyanate—toluene diisocyanate (TDI).

The isocyanate index, which controls the hardness of the foam, ranges from 100–120. In foams that use water as a blowing agent, water creates polyurea in the polyurethane that, in turn, increases hardness in the polymer.

8.2 MECHANICALLY FROTHED POLYURETHANE

Mechanically frothed polyurethane uses equipment that is similar to latex-coating equipment. The mixing equipment has additional metering pumps to move the components of the polyurethane to an "Oakes" type mixer (Fig. 8.1). Air or inert gas is used as the foaming agent.

The mixer must have sufficient cooling capacity to remove heat from the mixing head. The heat inside the mixer is generated from the shearing action of the materials. A creamy froth from the mixer is applied to the back of the carpet and is spread with a conventional knife or roll coater. The carpet is passed under an infrared heater through a convection oven set for a curing temperature range from 250°–300°F or over steam plates to react the polyurethane.

Figure 8.1. The Oakes mixer used for latex production. *(Courtesy http://www.continuousmixer.com/chem-latx.htm)*

This process uses a delayed-action catalyst which prevents the polyurethane from reacting at room temperature for a period of time. Examples of this type of catalyst are dibuytltindilaurate and acid salts of tertiary amines. Since the polyurethane reacts and cures without chemical blowing, polyurea hard segments are not formed. The polyurea segments are formed when water reacts with the isocyanate. Stiffening of the foam occurs from increasing the covalent cross-link density, use of polymer polyols, and inorganic fillers. Mechanically frothed polyurethane makes a stable-formed foam. A re-gauging roll can be used at the end of the heating cycle to help maintain uniform thickness of the foam.

8.3 WATER-BLOWN POLYURETHANE

The water-blown polyurethane method uses a highly reactive polyurethane system to create a low-density foam. This system can be used for a polyurethane precoat and foam single application or a polyurethane coating over a precoat. The foam application is adjusted by the amount of foam put on the carpet, the speed of the carpet-coating range, and the pumping speed of the mixer. An embossing roll or re-gauging roll at the end of the process can be used to maintain uniform thickness of the foam.

This process is different from the mechanically frothed process due to the abrasion-resistant skin that forms on the outer surface of the foam. The foam is blown and stiffened by the reaction between water and isocyanate. The low density of this polyurethane has greater load-bearing characteristics which creates walking comfort compared to other polyurethane foam systems. A disadvantage of this system is the possibility that defects from the foam application to the carpet will show on the back of the carpet. This can be hidden somewhat with the use of an embossing roller.

9

Cushion

Carpet cushion is used in residential and commercial settings. In either case, the foam can be attached to the carpet or be a separate pad. The foam primarily used for both residential and commercial broadloom carpet is made of polyurethane.

9.1 RESIDENTIAL

An industry-wide standard based on any single performance characteristic is not possible since there are other types of foams on the market. Some general guidelines for housing have been set by the U.S. Department of Housing and Urban Development (HUD). These specifications indicate which cushion application should perform well and help maintain the face appearance of carpet.

Selection of the carpet cushion is based on three main factors, traffic, location, and desired feel. *Traffic* is how much on which the carpet is walked; the more foot traffic, the firmer and thinner the cushion. For locations such as, hallways, stairs, and the most widely walked- in rooms, thin and heavy cushion should be used. In areas that are used for relaxation and where a cushiony feel is desired, a more resilient and thicker cushion should be applied.

A general guideline for polyurethane foam is discussed in Table 9.1. A discussion of polyurethane foam in general follows.

For residential foam, a high load-bearing foam is typically used. Due to environmental concerns, water is used as a blowing agent to make this foam. The density range of the cushion can range from 1 to 8-lbs/cu ft. Most prime carpet pads are made in the 1.5 to 3-lbs/cu ft range.

Over the past few years, the use of prime pad has been declining in favor of rebond or bonded foam. The reason for the decline is the belief that the durability of the prime carpet pad is not as good as the rebond. Therefore, the appearance retention of carpet would suffer if prime pad was used in an installation. Rebond pad has a higher density than the prime pad, which is believed to decrease the deterioration of the foam.

Most residential carpets do not have an attached cushion. Latex-coated carpet is widely used in this application. If polyurethane foam is applied directly to the carpet, it is applied to the back of the latex-coated carpet, heated and cured. The typical density of the polyurethane in this application method is 2.5–3 lbs/cu ft.

Table 9.1. Guideline Table for Cushion

Foot Traffic	**Light to Moderate**		**Heavy**	
Cushion	**Density (lbs/ft^3)**	**Thickness (inches)**	**Density (lbs/ft^3)**	**Thickness (inches)**
Prime Foam	2.2–2.7	0.25–0.375	2.7 plus	0.25 or less
Rebond	5.0 plus	0.375 plus	6.5 plus	0.375 plus
Frothed	10	0.25	12	0.25

9.2 COMMERCIAL

In the commercial carpet market, a denser foam is required for performance. Instead of using chemically-blown polyurethane, a mechanically air- or inert gas-frothed foam is used to produce a denser foam. In addition, calcium carbonate and aluminum trihydrate are used to reduce cost and provide flame retardancy. The majority of carpets used in commercial applications has an attached cushion. The carpet would have either a polyurethane or latex coating applied to the back of the carpet first.

End uses for polyurethane foam are normally broken into three categories. These are low, medium, and high foot traffic areas. Low traffic areas would have up to 500 foot traffics per day, medium traffic areas would have 500–1000 foot traffics per day and high traffic areas would have 1000 plus foot traffics per day. The foam density recommended for each application would be 8 to 12 lbs/cu ft for low, 12 to 16 lbs/cu ft for medium, and 16 lbs/cu ft and higher for high foot traffics.

These densities are not the actual density of the polyurethane itself. The density is calculated to include both the polyurethane and filler. For example, if one subtracted out the filler in 18-lbs/cu ft foam, the density of the polyurethane foam by itself could be 7-lbs/cu ft.

The polyurethane cushions used in the carpet industry are very similar from one company to another. Dow Chemical Company supplies most of the polyurethane chemicals. Most of the technology for producing the polyurethane foam is provided by Dow Chemical Company or Textile Rubber.

9.3 FOAM PERFORMANCE

The use of latex foam cushion has decreased significantly over the years. This is a result of their poor durability. Polyurethane foams are more durable and, therefore, are widely used today.

Besides density, some test methods used to compare performance of different cushions are compression set and load deflection. In the compression set test, foam is measured for total thickness and then compressed according to the customer's or supplier's specified time and temperature. Once the compression is released, the foam thickness is measured again. The result is expressed as a percentage loss of the original thickness.

In a load deflection test, the load-bearing ability of the foam is measured. A preset deflection is chosen and the weight in pounds required to reach the deflection point provides the value. This test method, which is no longer used as a description of load-bearing ability, is ASTM Method D1564.

9.4 CUSHIONS AND PADS

Carpet cushions made of felt, rubber and polyurethane are some of the pads being sold in the United States. Various densities, thicknesses, and weights are available depending on the application. Three general construction methods for pads are described here.

- **Needled Felt.** Needlepunch processing is typically used to make felt pads. These pads may contain carpet trim or waste, and/or a combination of hair and other fibers. Because of the contents, felt pads may cause problems with people who have allergies. These pads may mildew on concrete and may not be dimensionally stable.
- **Latex or Rubber Felt.** Latex or rubber felt cushion is made by coating a polyurethane pad with latex or rubber. One or both sides may be coated. This coating helps prevent moisture absorption and increases the dimensional stability.
- **Foam Rubber.**Natural and synthetic rubber is formed into a sheet. Typically, glues are applied to one side of the cushion for installation purposes. These types of foams can be made into different patterns. While the pads are resilient, they can be damaged by cleaning agents.

The HUD Assistance Secretary for Housing Commissioner sets minimum standards, expressed as classes, for carpet cushions.

- Class 1 is for light and moderate foot traffic such as living rooms, dining rooms, and bedrooms.
- Class 2 is for heavy foot traffic areas such as lobbies and corridors.

Cushions and their class standards are listed in Table 9.2.

Table 9.2. HUD Standards for Carpet Cushions

Type of Cushion	Class 1		Class 2	
	Thickness (inches)	Density (lbs/ft^3)	Thickness (inches)	Density (lbs/ft^3)
Flat Rubber	0.22	18	0.22	21
Ripple Rubber	0.285	14	0.33	16
Rubberized Hair Jute	0.27	12.3	0.375	11.1
Synthetic Fibers	0.25	6.5	0.3	6.5
Resinated Recycled Fibers	0.25	7.3	0.3	7.3
Bonded	0.375	5	0.375	6.5
Densified Prime	0.375	2.2	Not recommended	
Grafted Prime	0.375	2.7	0.25	2.7
Prime Urethane	0.375	2.2	Not recommended	
Mechanical Froth	0.25	10	0.25	12

10

Polyvinyl Chloride Plastisol Coating

Polyvinyl chloride (PVC) plastisol coating is used almost exclusively in carpet tile or six-foot-wide carpet construction. Carpet tiles are processed on equipment that can fuse a plastisol to a solid. Gas-fired infrared ovens or convention ovens can be used.

10.1 RAW MATERIALS

A plastisol is a liquid system containing PVC resin dispersed in a plasticizer. The viscosity of a plastisol compound can range from very low to very high. Plastisols are used to coat carpet as a liquid. The resin swells and is dissolved in the plasticizer when the plastisol is heated. Fillers and blending resins can be added to reduce cost. Other additives are used depending on applications.

Polyvinyl chloride carpet-backing materials must be selected to form a compound that will fuse at a temperature that does not have a detrimental effect on the carpet being produced. Therefore, a low fusion vinyl-chloride/vinyl-acetate copolymer dispersion resin is typically used. A low fusion temperature phthalate plasticizer is used at a level of 60–100 phr.

Resins should be chosen that produce a high yield along with a low viscosity under high shear. Rapid gelation and fusion are preferred to improve processing speeds.

10.1.1 Dispersion Resin

Dispersion resins are fine particles of PVC. These materials are produced from emulsion polymerization. Resins are classified into three areas: polymer type (homopolymer or copolymer), molecular weight, and particle size and shape.

A copolymer is PVC resin that has been blended with a non-vinyl chloride monomer. Most common copolymers contain between 3 and 7% vinyl acetate. Copolymers are used in applications requiring low-fusion temperature or improved adhesion.

High-molecular weight resins are used in applications requiring high physical properties. They require higher fusion temperatures and times to obtain these properties.

Size and shape of the resin particles contribute to the viscosity characteristics of a plastisol. The particle size of a dispersion resin is approximately 0.5 to 2.0 microns. Small particle sizes and irregular shapes increase the resin surface area, which increases plasticizer absorption and plastisol viscosity.

10.1.2 Blending Resin

Blending resins are small particle-size suspension resins. The blending resins range in size from approximately 10 to 150 microns. Blending resins are added to plastisols to lower cost and viscosity. They produce a matte finish on the surface of a product. Blending resins can settle out of a plastisol compound, particularly when used in high concentrations in a low-viscosity compound.

10.1.3 Plasticizer

Plasticizers are general colorless liquids which are relatively non-volatile. Most of the plasticizers used for carpets are esters of phthalic acids (phthalates) with a wide variety of long chain alcohols. Plasticizers are classified according to function or structure. The functional classifications are either primary or secondary. Plasticizers are used to convert PVC, a rigid plastic, into a soft, flexible, and elastic material.

A plasticizer that is compatible with PVC and exhibits low volatility, good permanence, and high efficiency is referred to as a *primary plasticizer*. A *secondary plasticizer* has the opposite characteristics of a primary plasticizer. The amount of plasticizer needed to obtain a certain hardness or elongation is a measure of its efficiency. The most efficient plasticizers are the most highly solvating and fastest fusing.

10.1.4 Stabilizers

Heat stabilizers are generally added to all plastisol formulations. Stabilizers have the ability to intercept hydrochloride from PVC decomposition and preventing the PVC from further degradation. Common stabilizers are made from barium-cadmium-zinc, cadmium-zinc, or octyl-tin compounds.

10.1.5 Thixotropic Agents

Thixotropic agents or thickeners are fillers that impart a high pseudoplastic ratio to the plastisol. They increase the low shear viscosity with minimal effects on the high shear viscosity of the plastisol.

10.1.6 Surfactants

Surfactants are added to plastisols to reduce viscosity, and improve viscosity aging , and air release. Polyethylene glycol derivatives are effective surfactants in plastisols. Adding 0.5 to 1.5 phr to the plastisol can reduce the viscosity by 10% to 20%.

10.1.7 Pigments

Pigments are typically carbon blacks or oxides of metal. Dispersing solid pigments in the plasticizer prior to adding them to the plastisol can improve dispersion in the compound.

10.1.8 Fillers

Fillers are added to reduce cost, improve insulation, and improve scuff resistance. At high-filler loadings, plastisol viscosity can rise significantly. The most common filler is calcium carbonate.

10.1.9 Lubricants

Generally, mixtures of waxes and fatty acids having average carbon chain lengths of C_{12}–C_{14} can be used as lubricants. They can reduce the tackiness of coated materials.

10.1.10 Blowing Agents

Chemical blowing agents are used to produce vinyl foam for improved foot comfort under the carpet. Proper selection of the blowing agent is needed to produce the desired density and cell formation of the foam. In order to produce closed cell foam, the plastisol should be formulated so the blowing agent decomposes at the point of the plastisol turning to a solid, but before the full properties of the PVC are formed. To do this, resin molecular weight, plastisol gelation temperatures, and blowing agent decomposition temperature must be balanced. To produce open cell foam, the blowing agent should decompose near the gelation temperature of the plastisol.

10.1.11 Solvents or Diluents

Solvents and diluents are added at levels of 5% or less to the plastisol to reduce viscosity. Two common solvents are methyl ethyl ketone and isobutyl acetate. Toluene and xylene are examples of diluents.

10.2 TROUBLESHOOTING

Typical problems with plastisol compounds are listed in Table 10.1 along with possible solutions. Table 10.2 gives typical problems that occur or flaws that appear in finished carpet.

Table 10.1. Troubleshooting Plastisol Compounds

Problem	Solutions
Compound is poorly dispersed	Increase mixer shear by withholding some liquid until solids are dispersed by increasing mixer speed by adding filler after resin is wetted out by filler absorbing plasticizer quicker than PVC by increasing mixing time
Viscosity is too high	Use a less solvating plasticizer as part of the plasticizer system Increase plasticizer level Add surfactant or diluent to compound Decrease filler level or change to a filler that absorbs less plasticizer Increase blending resin or use a coarser blending resin Change from a copolymer to a homopolymer Use a coarser particle size dispersion resin Check to see if compound is gelling during mixing Check to see if compound dispersed evenly
Viscosity is too low	Reverse steps in "Viscosity is too high" section Add thixotropic agents
Viscosity is inconsistent between batches	Keep plastisol and raw materials at the same temperature for mixing and storage Measure viscosity at same time intervals Make sure mixing procedures are followed
Plastisol gels during mixing or storage	Keep plastisol below 90°F Check copolymers as they are more sensitive to storage temperatures
Gelation temperature is too high	Use a more solvating plasticizer Use a lower molecular weight resin Use a copolymer or higher vinyl acetate level copolymer Decrease plasticizer level
Gelation temperature is too low	Use a more solvating plasticizer Use a lower molecular weight resin Use a copolymer or higher vinyl acetate level copolymer Decrease plasticizer level Reverse steps in "Gelation temperature is too high"

(Cont'd.)

Table 10.1. *(Cont'd.)*

Problem	Solutions
Coarse particles in compound	Use smaller size blending resin or filler Pre-disperse any solid components
Poor heat stability	Check age of stabilizer Increase stabilizer level Change stabilizer system Change from copolymer to homopolymer
Poor air release	Decrease plastisol viscosity Change to a resin with better air release

Table 10.2. Troubleshooting Finished Carpet or Processing

Problem	Solutions
PVC burns or degrades in oven	Raise fusion temperature and time Use higher molecular weight resin Decrease filler loading Decrease plasticizer level Change plasticizer system Check temperature controls and line speed for problems Decrease processing time or temperature
Bubbles in PVC	Check de-aeration equipment for leaks Increase de-aeration time De-aerate PVC Increase plasticizer level Use a copolymer instead of a homopolymer Use a more efficient plasticizer
Streaks in the PVC	Mix may be poorly dispersed Check for grind problems
Poor adhesion to carpet	Decrease plastisol viscosity Decrease gelation temperature Increase temperature at the PVC application Increase pressure at the PVC application Decrease temperature of pre-heater

(Cont'd.)

Table 10.2. *(Cont'd.)*

Problem	Solutions
Penetration of PVC through carpet face	Increase plastisol viscosity Decrease gelation temperature Increase temperature of pre-heater
Backing has poor gloss	Look at PVC grind Raise fusion time or temperature Evaluate chill or embossing roller
Backing is too hard	Increase plasticizer level Use a more efficient plasticizer
Backing is too soft	Reduce plasticizer level Use a less efficient plasticizer
Poor flammability results	Add ATH, antimony trioxide, phosphates, or chlorinated plasticizers to the PVC
Plasticizer migration	Use a more permanent plasticizer Decrease secondary plasticizer amount Use a homopolymer in place of a copolymer
Problems with foam plastisol	Check on compounding ingredients and refer back to discussion on compounding Check temperatures and line speed

10.3 FORMULATION

A typical PVC formulation is shown in Table 10.3.

10.4 DIMENSIONAL STABILITY

PVC used in carpet backing is described as a plastisol. Plastisols are a generic name for PVC particles dispersed in plasticizers that, when heated, fuse together to form a tough plastic film.

Table 10.3. Typical PVC Formulation

Material	PHR (Parts per Hundred Resin)
PVC Dispersion Resin	60–100
PVC Blending Resin	0–40
Plasticizers	30–100
Filler	50–250
Stabilizer	0.5–5
Pigments	0–5
Lubricants	0–2
Solvents	0–10
Other Additives	0–5

The advantages of these PVC plastisols are that they are applied at room temperature and do not contain water and solvents. Any volatility is dependent on the plasticizer used. If the correct plasticizer is used, there is no shrinkage. The dimensional stability, softening, or hardening are controlled by the plasticizer content and the molecular weight of the PVC. Polyvinyl chloride does not suffer from warm flow on heating up 140°F. When temperatures are used to heat PVC to 140°F and then reduced again to 0°F, the PVC does not become brittle.

Carpet backed with PVC is normally combined with fiberglass scrim. This fiberglass scrim supplies properties needed for dimensional stability. Glass fibers have a low coefficient of thermal expansion and further stiffen the PVC plastisol.

10.5 FLEXIBILITY

The inherent flexibility means PVC may be applied to a carpet and rolled. The rolled carpet does not have memory under tension which would give a dome or curl effect when the carpet is unrolled. Polyvinyl chloride reduces the effect of curling or doming because it is much higher in density (1.5 times) than other petroleum-based compounds and much heavier than alternative film thickness (1.75 times).

10.6 INDENTATION

Static loads affect PVC plastisols very little over a wide range of temperatures. Hot melt systems have high indentation at high ambient temperatures. Static load performance can be an excellent indicator of long-term performance of the carpet.

10.7 CUTTING CARPET INTO TILE

Carpet backed with PVC plastisols are easily cut into tiles. The die blades do not have a sticky residue on them as they would with a hot melt backed carpet.

10.8 IMPERMEABILITY

PVC is one of the best plastics known in resisting water and water vapor. This attribute makes it an acceptable backing in areas of health care. Refer to Ch. 14, "Antimicrobials," for more information on carpeting in the health care industry.

10.9 SUMMARY

Fiberglass reinforcing sheets are used in PVC-backed carpet tiles. The fiberglass prevents creep of the PVC. The crystallinity in PVC is reduced with the amount of copolymer resins used in the formulation. Resins and plasticizer suppliers supply most of the beginning formulations for manufacturers.

11

Hot Melt Coating

Hot melt adhesives have been used for years as an adhesive replacement for latex. Hot melts are thermoplastic materials that can be easily spread or poured when heated. When the hot melt is in a fluid state, it flows into the yarn on the back of tufted carpet. When the hot melt is cooled, the coating solidifies and forms a bond to the carpet. This same hot melt composition can be used to laminate a secondary backing to the carpet.

Hot melt-coated carpet is used when high performance requirements are needed in an installation. Due to its high equipment and raw material cost, it does not economically compete with latex-coated carpet. Hot melts typically out perform latex-coated carpets and do not lose their strength when exposed to moisture.

11.1 DISCUSSION OF HOT MELTS

Rapid-bonding characteristics are produced from this coating method. Hot melts have excellent moisture resistance and form good bonds at surfaces with little penetration. Heat resistance is poor, however, particularly with some ethylene vinyl acetate (EVA) formulations.

Most hot melts used in the carpet industry contain a combination of EVA, tackifiers, wax, oil, and mineral filler. Ethylene vinyl acetate is used because it is a copolymer of ethylene and vinyl acetate. These two components of EVA allow a formulator and processor of hot melts flexibility in finished compounds. When the percentage of vinyl acetate (VA) increases, the melting and freezing points of the EVA decrease. The hardness and tensile strength are reduced at the same time. This is due to the crystallinity of the copolymer. The increase in vinyl acetate increases the flexibility, adhesion, and elongation of the compound. Solubility in organic solvents is improved with increasing vinyl acetate content. A low-VA content EVA will have a higher degree of crystallinity than a high-VA content. The melting point predicts high temperature performance and the set speed is determined by the freezing point.

Ethylene vinyl acetate copolymers can thermally cross link at 190°C. This may give a low indication of the flow ratc. Often, a lower extrusion temperature of 125°C is used to determine melt flow rate (MFR). ASTM Test Method D1238 is used to determine MFR.

Melt index (MI) is a key parameter in hot melt compounds. The MI is an indication of viscosity of a compound, and it is found using an extrusion plastomer or melt index tester. Melt index is measured by the grams of a compound that comes out of a narrow orifice held at 190°C with a specified weight over a ten second period of time using ASTM D1238. A high MI indicates the compound viscosity and molecular weight are low, and the reverse is true of a compound with a low MI.

The primary use of melt index (MI) is for coating properties of the compound. Lowering the melt index of a compound increases the softening point, elongation, stiffness, and hardness of the hot melt. The end use temperature for a hot melt is the deciding factor in its use. Since the addition of other components contributes to the final performance of a specific compound, melting points must be known because creep or failure will occur if the hot melt becomes a liquid again. .

The viscosity of latex used in carpet coating is usually measured with a Brookfield viscometer. Ethylene vinyl acetate copolymers and hot melt compounds have too high of a viscosity to be measured this way. The following formula can be used to convert MI to viscosity at 190°C.

Eq. (11.1) Viscosity, cp at 190°C = 8,400,000/MI

11.2 INGREDIENTS

Hot melt compounds must have the correct components to achieve maximum performance and manufacturability. These compounds can contain polymers, resins, waxes, stabilizers, fillers, and oils.

11.2.1 Polymers

Polymers are selected for their adhesion to components of carpet. To achieve the best possible adhesion to the carpet, a polymer must be chosen based on the following. The polymer structure must bond the structure together, which is determined by the backbone of the polymer. Flexibility, strength, easy viscosity adjustment, solubility of the polymer with resin, and melt strength must be also be considered.

11.2.2 Resins

Resins or tackifiers should be chosen on the basis of solubility with the polymer to achieve maximum adhesion. Odor, heat stability, and compatibility of the resin at high temperatures must be taken into account when formulating a hot melt. For carpet applications, amorphous low molecular weight resins are used to modify the polymer.

11.2.3 Wax

Waxes are used in a hot melt compound to lower viscosity. At the same time, waxes control the open time and setting speed of the compound. *Open time* is the time to form a bond between applications of the hot melts to the carpet. Typically there are three types of waxes. These are paraffin, micro, and synthetic. Melting temperatures of these waxes are 200°F plus, 155°–195°F, and 120°–155°F, respectively.

11.2.4 Filler

Mineral fillers are low cost additives that replace the higher cost polymers and resins. Fillers improve the abrasive resistance of hot melts. Calcium carbonate, in concentrations of 30%–60%, is typically used to lower the cost of the hot melt compound and to increase its mass. When barium sulfate is the filler, a higher concentration is used.

11.2.5 Oils

Oils are used as processing aids in coatings or as modifiers to the finished hot melt compound. Oils improve the flexibility, resiliency, and melt flow of the coating by acting as a lubricant between molecules. Raising the vinyl acetate concentration can increase the solubility of the oil into a hot melt containing EVA.

11.2.6 Antioxidants

Oxidation occurs when oxygen reacts and removes hydrogen atoms on the hot melt molecules. Antioxidants in powder form are used to prevent the thermal degradation or oxidation of the hot melt. The removal of the hydrogen causes reactive sites, i.e., free radicals, to form in the compound, which weakens the composition. They are typically in fractions of 1% of the overall compound formulation.

11.2.7 Flame Retardants

The same explanation of flame retardancy used for carpet with a latex coating (Ch. 7, "Latex Coatings," Sec. 5) can be used for hot melts also. As with latex coatings, aluminum trihydrate (ATH) is the flame retardant of choice for hot melt compounds.

11.3 HOT MELT COATING PROCESS

Hot melt compounds are generally applied by processing tufted carpet over two applicator rollers. These application rollers have another set of rollers above them, called press rollers, to control the contact pressure of the carpet to the application roller, and a doctor bar on the roller itself, to control the amount of hot melt picked up by the roller as it comes out of the hot melt. Both applicator rollers are positioned in containers or reservoirs of hot melt.

In a 100% hot melt-backed carpet, two types of hot melt compounds are used. The first compound penetrates the fiber on the back of the carpet. The material must have a low viscosity in order to penetrate the yarn, which serves the same purpose as latex in an aqueous coating. This low viscosity compound can be tackifiers and/or EVA with little to no filler. Tackifiers can be described as low molecular weight hydrocarbons. These hydrocarbons can be aliphatic, aromatic, or a combination of the two. The first coating wets out the fiber so the more viscous second coating can penetrate and bond to the yarn also. The first coating's weight is generally 6–12 oz/sq yd and the second coating's weight can be approximately 24–34 oz/sq yd. Some hot melt-backed carpet uses carpet which has been previously been coated with latex. Therefore, the first hot melt coating application will not be needed.

Applicator rollers are used in hot melt coating because they are one of the most versatile methods of applying materials to a carpet. A wide range of viscosities can be used in the reservoir pan of the hot melt. Adjustments to coating weights can be made by adjusting the pressure on the carpet above the roller, the roller speeds, and the doctor bar.

The second sheet of hot melt can be produced off-line and then laminated to the first molten hot melt coating. This eliminates the need for two hot melt applicators and allows for flexibility in the formulation and performance of the final backing.

In general, the second coating of hot melt contains a higher level of filler, lower levels of tackifiers, and higher concentrations of EVA. While a wide range of VA content in EVAs can be used, hot melt compounds for carpet application have a VA content of 18%–28%. If the VA content is below 18%, the compound stiffens. Above 28% VA content, the compound becomes easy to elongate and can become oily if an oil has been added to the compound. Oils of a napthenic, paraffinic, or aromatic structure are used in a 1%–10% by weight content. High oil content causes the melt index of the final compound to rise rapidly as the concentration increases. High oil concentrations increase the chance of the oil leaching out of the finished carpet over time.

The melt index of EVA in the second hot melt composition can range from approximately 1 to 50. As a general rule, EVAs with a melt index between 1 and 30 are used to meet the strength requirements of the compound. In the final hot melt compound, the total EVA by weight is approximately 10%–30%.

11.4 COMPOUNDING

A quality homogenous composition must be made for a hot melt compound. Basic mixing procedures must be followed. The surface area of the components must be dispersed so all materials come in contact with each other. A high-intensity mixer such as a Banbury high intensity mixer or Farrel continuous mixer (FCM) can be used to compound the hot melt.

Resins or tackifiers are added into the compounder. The solid material becomes molten when heated. Once the material is molten and is homogenous, oil can be added. The filler is added to the melted resin slowly. The mixture must be maintained above 300°F while the filler is being added. Inadequate mixing or mixer shut down will occur if this temperature is not maintained. Once the mixture is homogenous again, polymers are added to the mixture. Polymers may be added to the hot melt compound earlier if the compound will be highly filled.

The finished compound is held in a molten state if the compounder will use it. If the hot melt compound is to be used at another location, it is pumped into a drum to be remelted by the user or shipped in a molten form to the end user.

11.5 EXAMPLES OF HOT MELT COATING COMPOUNDS

Table 11.1 has examples of hot melt coating compounds and their application. One is for broadloom carpet and the other for tile. Each formula is based on specific final properties desired by the carpet manufacturer, since it is difficult to use a single compound to meet all physical and performance properties.

Along with the information provided in this chapter, more information about hot melts and their components can be found in literature written by Equistar Chemicals and Hercules Incorporated.

Table 11.1. Examples of Hot Melt Compounds

Broadloom with Secondary		
First Application	6–12 oz/sq yd	Tackifier
Second Application	18–24 oz/sq yd	Compound consisting of: 25% EVA 25% Tackifier 50% Filler
Secondary applied to the carpet after the second application.		
Broadloom with Unitary		
First Application	6–12 oz/sq yd	Tackifier
Second Application	18–24 oz/sq yd	Compound consisting of: 12.5% EVA with low VA 12.5% EVA with high VA 25% Tackifier 50% Filler
Carpet Tile		
First Application	6–12 oz/sq yd	Tackifier
Second Application	18–24 oz/sq yd	Compound consisting of: 35% EVA 35% Tackifier 30% Filler
Fiberglass applied to carpet after the second application		
Third Application	30–50 oz/sq yd	Compound consisting of: 20% EVA 20% Tackifier 60% Filler

12

Extrusion Coating Technology

Extrusion coating is a process which forces a thermoplastic material through a die to form a sheet. Extrusion coating technology has been known for years. Only recently has the commercialization of this process for coating carpet with thermoplastic polymers occurred. The thermoplastic polymers replace latex as part of or the entire adhesive for carpet.

Many of the same technical aspects, principles, and compounds used in hot melt coating can be applied in extrusion coating. The most significant difference between hot melt coating and extrusion coating is the viscosity of the material used.

12.1 EXTRUSION COATING PROCESS

The extrusion process is divided into three basic parts. These consist of an extruder to melt the plastic, a die to form the plastic, and downstream equipment. The downstream equipment is needed to maintain the plastic's final shape.

Carpet is coated through an extrusion process similar to latex or hot melt coating. The major difference is the method in which a coating is applied to or another material is laminated to the carpet.

An extruder melts the compound and the compound is transferred to a die which applies it across the width of the carpet. When melted plastic exits the die, it is called extrudate. The melted plastic, under pressure while in the die, expands when it exits the die. This expansion is called die swell. The amount of die swell is dependant upon the materials in the plastic and the temperature of the melt.

The extrudate is applied to the back of the carpet and penetrates the yarn. If the carpet already has another type of coating applied to it, the extrudate must bond to the coating rapidly because the extrudate cools quickly.

12.2 THE EXTRUDER

The extruder is a melt supply unit in the extrusion process. Molten material is supplied at a uniform and constant rate. While most extruders used to coat carpet are single-screw extruders, the use of twin-screw extruders is possible.

For specifics on the extrusion process, extrusion design criteria, the functional and geometric zones, and mixing structures on the extruder screw, refer to the literature. The *Plastic Extrusion Technology Handbook* by S. Levy and J. Carley, in particular, may be helpful.

12.3 EXTRUDER DIE

A sheet die is attached to the end of an extruder either directly or with an adaptor. Sheet dies are used when varying thicknesses are required. The die opening needs to be adjustable to control the flow in different parts of the die in order to control the thickness across the sheet width. However, most of the control of the flow and thickness needs to be made prior to the polymer reaching the die lips.

Two types of dies are used for applying a sheet of plastic onto the back of carpet with an extruder. These are a T-die and a coat hanger die. Both die configurations consist of a large feed channel that directs the melted plastic to the back edge of the deck surface. Reduction of flow and equalization of pressure of the plastic in the die takes place on the deck. As the pressure increases behind the deck, the plastic is forced over the deck and out of the die.

With the T-type die, the plastic has to travel a greater length on the preland section to fill the end of the "T". This reduces pressure at the ends and increases the amount of residence time the plastic is in the die feed channel. The coat hanger type of die overcomes this problem by extending the middle section of the die in the shape of a coat hanger. The triangular preland area inside the die results in a large pressure drop on the preland area near the feed point of the die, and less of a pressure drop at the end of the die. Selecting the proper angle allows a well balanced flow of polymer to the die lips. This minimizes the adjustments required at the die lips to maintain uniform coatings. This construction equalizes the distance the plastic has to travel along the entire length of the preland section. A properly designed coat hanger die allows plastic to have the same residence time and the same pressure drop while in the die.

High operating pressures require the die to have varying sizes, shapes, and number of bolts to hold the two die manifolds together. For extrusion coating carpet, a backpressure of approximately 500 to 2,000 pounds per square inch (psi) will be created.

12.4 DOWNSTREAM EQUIPMENT

Solidification and forming of the coating is quickly achieved once the sheet comes in contact with the tufted carpet and passes between a press and a chilled roller. The amount of plastic compound applied to the carpet is controlled by the gap in the die lips and the gap between the press and chilled rollers. Ambient air or fans can achieve total solidification. A press roller is used to apply force against the carpet so the thermoplastic polymer must be in a molten state to encapsulate the yarn on the back of tufted carpet. Penetration and adhesion of the molten polymer is determined by temperature, pressure, and time.

Other types of materials, such as cushions, reinforcements, or secondary backings, can be laminated to carpet using the extrusion coating process. These materials are positioned on the opposite side of the die. The extrudate is sandwiched between the carpet and the material to which it is being bonded.

The chilled roller is actually a shell which is cooled by circulating water inside the hollowed area. The chilled roller and press roller can help pull the polymer out of the die. A gap between the press roll and the chilled roller does provide some control of the final thickness of the coating.

12.5 POLYMERS AND COMPOUNDS

Extrusion coating can be used with polymers and with polymer/additive compounds. The thermoplastic polymers typically are composed of polypropylene, polyethylene, polybutylene, and thermoplastic elastomers. The temperature of the molten polymer is between 300°–700°F. The polymer coating method does not require a drying step, as does the latex coating method. A secondary coating can be applied using the method.

Typically, the direction of the extrusion coating of the carpet is horizontal. The molten material is forced out of the die by the extruder and by its own weight. Shear and tensile stresses on the polymer are lower from the gap of the die and its contact with the carpet than within the die. Pseudoplastic materials have high viscosity at low shear, therefore, shape change is minimized. An example of a highly pseudoplastic polymer material is poly(vinyl chloride); polyolefins are not. Materials that have non-Newtonian flow work more easily.

Crystallinity of the polymer influences the performance of the carpet because materials that are highly crystalline are difficult to control in cooling. An abrupt reduction in volume per unit mass accompanies the crystallization of the thermoplastic. The corresponding shrinkage in lineal dimensions is not always equal in all directions. If the polymer does not cool equally, the carpet backing distorts.

The viscosity and strength of thermoplastics can be affected by their heat history. Molecular degradation and reduction of physical properties can be caused by excessive heat history. Productivity of the extruded polymer may be affected.

12.6 EXAMPLES OF EXTRUSION COATING COMPOUNDS

Table 12.1 gives examples of extrusion coating compounds and how the polymers and additives affect the final properties of the compounds. The examples use different EVAs because it is difficult to use a single EVA in a compound to produce a satisfactory carpet backing. Ethylene vinyl acetate-based compounds typically require two EVAs and processing oil.

The processing oil is used to lower the temperature of the compound and improve flexibility of the backing. This flexibility is needed to prevent carpet backings from becoming brittle and stiff with high filler loadings. Ethylene vinyl acetates with high vinyl acetate concentrations are needed in a compound with a high oil content to prevent the oil from migrating out of the finished backing.

Table 12.1. Extrusion Compound Comparison*

Example 1			
Stiff Compound		**Flexible Compound**	
Components and Description	**Weight Percent**	**Components and Description**	**Weight Percent**
EVA (18% VA, 0.7 MI)	20	EVA (12% VA, 0.3 MI) EVA (40% VA, 57 MI)	16 4
Oil	7	Oil	7
Filler	73	Filler	73

Example 2			
Stiff Compound		**Flexible Compound**	
Components and Description	**Weight Percent**	**Components and Description**	**Weight Percent**
EVA (25% VA, 2 MI)	20	EVA (18% VA, 0.7 MI) EVA (40% VA, 57 MI)	14 6
Oil	7	Oil	7
Filler	73	Filler	73

* *Examples of compound comparisons taken from U.S. Patent Number 4,379,190 issued to E. I. DuPont de Nemours and Company.*

(Cont'd.)

Table 12.1. *(Cont'd.)*

Example 3			
Stiff Compound		**Flexible Compound**	
Components and Description	**Weight Percent**	**Components and Description**	**Weight Percent**
EVA (25% VA, 2 MI)	13	EVA (18% VA, 0.7 MI)	9
		EVA (40% VA, 57 MI)	4
Oil	7	Oil	7
Filler	80	Filler	80
* *Examples of compound comparisons taken from U.S. Patent Number 4,379,190 issued to E. I. DuPont de Nemours and Company.*			

It may appear from these compounds that the reason for the flexibility of the compounds on the right is the higher melt index of these compounds. This is not correct; the melt index of the two EVAs is approximately the same as the single EVA polymer. The flexibility is a result of the oil being absorbed more efficiently by the compound containing the higher vinyl acetate EVA.

13

Carpet Tile Coatings and Reinforcements

Carpet tile must have good dimensional stability to provide lay flat performance. The tile must be perfectly square for installation purposes and pattern matching; smooth and non-fraying edges are desirable.

Carpet tiles have been used for several years, primarily in areas that require high performance. Carpet tiles can be made from a wide variety of polymers and processes.

To produce a stable carpet tile, the carpet must be produced with the least amount of tension possible in all phases of production.

Carpet is coated and then a reinforcing fiberglass scrim (Sec. 13.2) or foam may be applied to the hot melt, PVC, asphalt, or extrudate. The coated material is applied in a fluid state and is solidified before the next coating. The fiberglass scrim is further coated with hot melt, PVC, asphalt, or extrudate, and the second coating is solidified.

13.1 COATING SYSTEMS FOR CARPET TILE

13.1.1 Polyurethane

The tufted carpet, which has a coating of latex, passes under the application roller, which applies atactic polyurethane. Right after the unreacted polyurethane saturates the carpet, a fiberglass reinforcing scrim is laid with the carpet. This composition is passed through a convection oven or over steam plates to react the polyurethane. After the reaction, a second coating of polyurethane is applied to the carpet. The reaction of the second application of polyurethane takes place in a second oven or over a second set of steam plates.

13.1.2 Polyvinyl Chloride (PVC)

If a carpet is to be coated with 100% polyvinyl chloride (PVC), the PVC precoat must have a low enough viscosity to penetrate into the carpet bundle. The second coating of PVC must be able to bond to the first PVC coating and penetrate the fiberglass that is applied to stabilize the carpet. The tufted carpet, which has a coating of latex or PVC, passes under an application roller, which applies a PVC plastisol to it. A fiberglass-reinforcing sheet or scrim is laminated to the liquid plastisol shortly after the carpet is coated with enough plastisol to saturate the carpet. The composition is passed through a convection oven or gas-fired infrared oven to fuse the PVC. After exiting the oven, a second coating of PVC is applied to the carpet. The fusion of the second coating of PVC takes place in a second oven.

13.1.3 Hot Melt

The same type of processing and raw materials used for broadloom carpet can be used for carpet tile. The tufted carpet passes over the first application roller where the bottom of the carpet surface is coated with the hot melt. A relatively fluid coating of a high melt index (MI) is applied in an amount to sufficiently bond the individual tufts. The composition is passed through a second hot melt applicator at which time a second application of hot melt is applied to the first coating. A reinforcing-fiberglass scrim or a reinforced backing with a fiberglass sheet is laminated to the carpet at the second coating. If the fiberglass reinforcement is applied by itself, the carpet will require another hot melt coating.

13.1.4 Extrusion

The tufted carpet passes under the extruder die where the bottom of the carpet surface is coated with the extrudate, which is a relatively fluid coating of a high MI. It is applied in an amount to sufficiently bond the individual tufts. The composition is passed through a press roller and a chilled roller at which time the secondary backing or reinforcing material is applied to the carpet. Instead of using a high MI coating to bond the yarn to the carpet backing, a latex coating can be substituted.

13.2 FIBERGLASS REINFORCEMENT

A fiberglass nonwoven material is the primary stabilizing layer used for carpet tiles. The fiberglass is inserted between the two backing layers. Fiberglass with a weight of 15–60 g/sq m is typically needed to stabilize carpet tiles. The position of the fiberglass should be between the following ratios of the top and bottom layers of the carpet, 1:10 to 1:2.

The fiberglass nonwoven material is needed to correct the unbalanced mechanical forces which occur during the production of the carpet. These differences are due to the strength of the length versus the width of the carpet. The backing weight of the carpet tile should be between 30–150 g/sq m.

The fiberglass should have some degree of coarseness and permeability to permit the backing composition layers to penetrate it. By doing this, the two layers will come in contact with each other and form good bond strength.

13.3 SUMMARY

Table 13.1 gives a comparison of the various coating systems.

Table 13.1. Comparison of Carpet Tile Coating Systems

Coating Material	Typical Wt. (oz/sq yd)	Processing Temp. (°F)	Advantages	Disadvantages
PVC	90–125	300–350	Stability, Mechanical Properties, Elasticity	Plasticizer, Smoke, Toxicity
APP	90–110	275–325	Stability, Stiffness	Flammability, Plasticity
Hot Melt	90–110	275–325	Stiffness, Tuftbind	Flammability, Stability

Part 4

Carpet Enhancers

14

Antimicrobial Agents

An *antimicrobial agent* is any variety of chemical or physical compound that can destroy or prevent the growth of microorganisms. These are a few examples of methods an antimicrobial agent can use to destroy microorganisms. Antimicrobial compounds can directly destroy the cells; they can penetrate the cell itself to prevent it from obtaining nutrients for survival, or for reproduction; and some coagulate materials from inside the cell, thereby, destroying the cell.

14.1 USE

Many institutions, such as hospitals, schools, airports, and office buildings, require the use of carpet for acoustic, comfort, and thermal properties; prevention of falls; and for other reasons. While meeting institutional requirements, carpet can be a host to microorganisms. Microorganisms can cause odors, can discolor and deteriorate carpet, and can cause infectious diseases. Vacuuming and shampooing do not remove microorganisms because of limited penetration of the pile. Environmental conditions play a role in how microorganisms multiply; microorganisms thrive in warm moist areas where they have food for survival.

The reasons for using antimicrobial agents in carpet are to control the spread of bacteria and fungi, prevent cross contamination, control odor, and to protect it from deterioration or discoloration. Carpets, because of the fiber used in its construction, can trap soil, dander, food, drinks, and pet excrement, which are nutrients for microorganism growth.

Many antimicrobial agents are very water soluble. If the applied antimicrobial is very water soluble, it will leach up to the surface of the pile. When the antimicrobial agent leaches, it can not protect the backing or the base of the pile. Some agents are only effective against specific microorganisms. With these, additional antimicrobial compounds must be added for full protection against a variety of microbes.

For an antimicrobial to be effective in carpet, it needs to have a broad spectrum of activity against numerous bacteria and fungi, yet have a low toxicity for animals; it must be stable during carpet processing; and be durable. Antimicrobial agents are compared by testing the activity against bacteria and fungi using the American Association of Textile Chemists and Colorists (AATCC) Test Method 174. This method is the test used under the General Services Administration (GSA) specification.

Carpet and its surrounding environment can be further protected by the addition of an antimicrobial agent to the adhesive holding the carpet to the floor. This prevents the adhesive from degrading should any moisture come in contact with the adhesive or if water seeps through the carpet.

Government contracts for carpet with antimicrobial protection can only be awarded to carpet manufacturers who produce carpet that meets the requirements set forth under the terms of GSA solicitation 3FNH-98-F301B. For a chemical to be advertised and used as an antimicrobial agent in carpet, it must be registered with the Environmental Protection Agency (EPA).

14.2 BACTERIAL SOURCES AND CONDITIONS FAVORABLE TO GROWTH

The human body provides the primary source of indoor bacteria. Other sources are heating, ventilating and air conditioning (HVAC) systems, toilets, humidifiers, and dishwashers. Contrary to popular belief, properly sanitized urinals and toilets are not a potential source. Bacteria and fungi are brought into the building from outside sources. The spread of infection is made possible by a person breathing in spores or bacteria of the infection and bringing them indoors in the respiratory tract. Dead or dying animals and plants provide sources of fungi. Coming in contact with or bringing these sources into the building quickly causes the spread of spores under favorable conditions.

Humidity levels above 70% are ideal for microorganism growth. Airborne pathogens can easily survive at this humidity level also. Microorganisms require a humidity level of 25% to 70% to multiply.

The combination of bacterial and/or fungal contamination and high humidity levels results in the development of an indoor environment supportive of the spread of respiratory diseases. Buildings with this indoor environment are said to have "sick building syndrome."

The prevention of microorganism sources is the best method of controlling sick building syndrome. This includes preventative maintenance: regular housekeeping, including the filters of air conditioners and air circulators; a reduction in humidity; and an increase in air turnover. Antimicrobial agents also can be useful in helping control bacterial and fungal growth.

14.3 ANTIMICROBIAL AGENT SELECTION AND CONSIDERATION

When a carpet manufacturer is considering an antimicrobial agent for use, humans, animals, and the environment must be taken into consideration. The antimicrobial agent needs to be screened to avoid volatile organic chemicals, toxic chemicals, allergens, skin sensitive material, photosensitivity or carcinogens. Some antimicrobials will cause carpet fibers to turn yellow, immediately or after the carpet has been installed, from UV light or oxides in air.

Antimicrobial agents may be applied in or on carpet fibers, the latex or adhesive holding the fibers, or the carpet backing. Carpet manufacturers must ensure that any topical antimicrobial agent is compatible with any other topical treatments such as stain blockers or fluorochemicals. For application in latex, consideration must be made for such things as pH or the antimicrobial agent leaching out of the compound once the latex is dry.

14.4 RESIDENTIAL APPLICATIONS

Microorganisms can cause odors in various parts of the house. They may be more susceptible to growth in areas of the house with high moisture content such as the kitchen or bathroom. Many antimicrobial suppliers have advertised products, i.e., many bacteriostats and fungistats for household use, which are very water soluble. Besides providing very little protection against microorganisms, they are easily removed when the carpet is cleaned. Some of these antimicrobials are only effective against a few microorganisms. Therefore, consumers need to become informed about selecting antimicrobial products which provide the maximum protection against microorganisms. Otherwise, consumers will lose faith in antimicrobials and their use in carpets.

14.5 PROOF OF CLAIMS

Many products have been marketed in recent years with various claims of pesticide treatment and implied health claims. Many of these make claims of protection against bacteria, fungi, and viruses or make claims against pathogenic organisms that might cause food poisoning, infectious diseases, or respiratory problems.

There is a difference between antimicrobial and antibacterial agents. *Antimicrobial* means that an article has been protected from attack by microorganisms whereas *antibacterial* means products are designed to control human pathogenic microorganisms.

The EPA has outlined acceptable terminology for claims which a product can make. Claims such as "mold or mildew resistant" are considered acceptable. This does not give any misleading impression that the product can protect users against disease-causing bacteria. These claims do not require EPA registration. Claims such as antimicrobial, bactericidal, etc., require registration as a pesticide product with an EPA registration number. Registration for an antimicrobial typically takes over a year.

Before a manufacturer can register an antimicrobial agent as a pesticide, the manufacturer must submit proof of the agent's effectiveness. The Federal Insecticide, Fungicide, and Rodenticide Act (FIFRA) regulates pesticides and requires that the pesticide be registered by EPA. By EPA definition, a *treated article* is an article finished with or containing a registered pesticide which provides protection to the article or substance. The manufacturer of the pesticide must deliver to EPA data from numerous tests indicating that there are no risks to humans or the environment.

14.6 MICROORGANISM STRUCTURE

To understand how an antimicrobial works, basic knowledge of three structures is needed. These are fungi and the two common types of bacteria—gram positive and gram negative. The microorganism's outer structures can determine the effectiveness of antimicrobials.

The smallest unit of life is a cell. A rigid wall protects the fluid membrane from external forces. Cell walls of bacteria and fungi are different. There is even a difference in cell walls of gram-positive and gram-negative bacteria. Fungi cell walls consist primarily of chitin, glucan, mannan, and diaminopimelic acid. Gram-positive bacteria have a rigid wall consisting mainly of peptidoglycan and teichoic acids. This protects the fluid cytoplasmic membrane beneath it. Penetration of small molecules such as antimicrobials is easy through the peptidoglycan. Gram-negative bacteria have a more complex cell membrane than gram-positive bacteria. Layers of lipoproteins, lipopolysaccharides, and other lipids cover peptidoglycan. This outer membrane prevents many molecules from penetrating into the cell wall. This has been frequently seen in testing of many antimicrobials.

The most important barrier between a cell and its environment is the cytoplasmic membrane. It is made of a double layer of lipid molecules. The double layer consists of a nonpolar hydrophobic hydrocarbon chain extended outward with hydrophilic terminal groups toward the aqueous medium.

A cell membrane is in a fluid state, which allows it to have flexibility because the bonds between the individual components are not covalent. This allows for diffusion of proteins within the membrane.

14.7 TYPES OF ANTIMICROBIAL AGENTS

There are several types of chemicals that can be used as antimicrobial agents. These can be divided into two basic groups; membrane-active compounds and electrophilic agents.

Membrane-active chemicals are adsorbed in the cytoplasmic membrane and prevent the electron transport chain and/or ATP synthesis. These chemicals include alcohols, quaternary ammonium compounds, biguanides, guanidines, acids, phenols, salicylanilides, and carbanilides.

All membrane-active compounds can be effective antimicrobials against bacteria. The concentrations of the antimicrobials are important for their effectiveness. These chemicals are adsorbed at the cytoplasmic membrane first. The more dissociated the antimicrobial, the stronger the absorption. Weak acids and phenols with low pH values and quaternary ammonium compounds and biguanides with high pH values have strong absorption into the cell. Hydrophobic properties of membrane-active chemicals are important in their efficacy.

The action of membrane-active substances can involve nonspecific adsorption at the cell membrane, disruption of the function of proteins, the escape of ions and organic molecules, and inhibition of substrate transport and ATP synthesis.

Electrophilic compounds react with nucleophilic components of the cell and inhibit activities within the cell which are vital for life processes. Some examples of electrophilic agents are aldehydes and substances that release aldehydes, activated halogen compounds, isothiazolines, and mercury compounds.

The strong electrophilic antimicrobials react with nucleophilic cell components and inactivate the enzymes and substrates necessary for life functions in the cell. Each chemical has a specific mechanism of action.

14.8 VARIOUS ANTIMICROBIAL TREATMENTS

Although different types of antimicrobials have been in use for centuries, in the past forty or so years, more consumers have become conscious of antimicrobial treatments for floor coverings. There have been several types of antimicrobials sold by different companies to the carpet industry to control bacteria and fungi.

From a chemical standpoint, there are three basic chemical compositions used in carpet applications. These are organo-phenols, organosilanes, and organometallics. They also can be combined with each other.

Some of the various types of antimicrobials that have been used in the carpet industry are bis(tri-n-butyltin) oxide, 5-chloro-2-(2,4-dicholorophenoxy) phenol (Triclosan, trichlorotrioxydiphenol), 10,10-oxybisphenoxarine (OBPA), 1, 2 benzisothizolin-3-one, 3-trimethoxy-silylpropyl dimethyloctadecyl ammonium chloride, or bis(1-hydroxy-2(1H)-pyridinethionato-O, S-(T-4) zinc (zinc pyrithione).

Morton International produced two antimicrobial products. They were Vinyzene® and Durotex®. The chemical composition was OBPA or a combination of OBPA and organometallic compounds. The OBPA material was a very effective broad spectrum antimicrobial agent used for years until public concern was raised about the arsenic component of its chemical structure; its use was reduced.

Sanitized® and Dow Corning produced quaternary silane-organosiliene compounds. Dow Corning products were called Sylgard®. Sylgard's active ingredient was 3-trimethoxylsilylpropyl dimethyloctadecyl ammonium chloride. A similar product from Sanitized was called Requat®. Dow Corning sells its antimicrobial agents such as Sylgard through AEGIS Environmental Management. Products are now sold under the AEGIS Microbe-Shield trademark. Clariant sells the Sanitized trademarked antimicrobial agent. Bioshield Technologies sells a quaternary silicone-based product using the Bioshield trademark. These products are typically applied topically to fibers. Interface Flooring markets a phosphated amine called Intersept® as an antimicrobial finish for its carpet.

Several companies have sold antimicrobials based on Ciba-Geigy's Irgasan DP300® (5-chloro-2-(2,4-dicholorophenoxy) phenol) and Irgagard®. These companies and the trademark names of their antimicrobials include Clariant® (Sanitized), Microban Products Company (Microban®), Thomson Research Associates (Ultra-Fresh®), and Vikon Chemicals (Vikol THP®).

Arch Chemicals manufactured products called Zinc Omadine® and Sodium Omadine®. The active ingredients are zinc pyrithione and sodium pyrithione, respectively.

An in depth discussion of all possible antimicrobials is not practical in this text due the microbiology involved. More information can be found in manufacturers' literature.

14.9 TESTING

Carpet is typically tested for antimicrobial inhibition using the American Association of Textile Chemists and Colorists (AATCC) Test Method 174. The test method contains three parts. Part I is a qualitative assessment of antimicrobial activity, Part II is a quantitative assessment of antimicrobial activity, and Part III is antifungal activity assessment. These test methods were basically derived from combining AATCC Test Methods 30, 100, and 147.

In Parts I and II, gram-positive and gram-negative bacteria are used. In Part III, only fungi are used. The terms gram-positive bacteria and gram-negative bacteria came from microbiologist Christian Gram. He discovered that one class of bacteria would stain with certain dyes and the other would not. Those bacteria which stained were known as gram-positive and the others as gram negative.

In testing of the efficacy of antimicrobials, it is important to try and select an organism that does not mutate easily. Normally, a pathogen (disease causing) is not desirable. This is not always possible. In AATCC Test Method 174, *Pseudomonas aeruginosa* is used as the gram-positive bacteria. This bacterium is a pus-forming organism that is found in moist areas of hospitals. It can be passed to different people from washcloths, hands, etc. It will mutate. For gram-negative, *Klebsiella pneumoniae* is used. These bacteria can be present when water has fecal contamination. *Aspergillus niger*, a common fungus in nature, is used for Part III.

In Part I of AATCC Test Method 174, a piece of carpet is placed into the bottom of a container holding hardened gel known as an agar. The agar contains nutrients for the bacteria to grow. Bacteria are placed onto the agar and the container is closed and subjected to a specified temperature and time. Petri dishes are normally used to easily observe bacteria resistance of the carpet. The resistance generally appears as a clear area around the carpet. This area is known as the zone of inhibition.

In AATCC Test Method 174, Part II, a 48 mm diameter circular disk of carpet is cut and inoculated with bacteria. The carpet is placed in a 250 ml glass jar and sealed. The jar is shaken and a bacteria count is taken. After 24 hours, the jar is opened and another bacteria count is taken. The results are compared, and the percent reduction in bacteria is calculated.

In AATCC Test Method 174, Part III, a similar test to Part I is conducted except the agar has nutrients suitable for fungal growth. A zone of inhibition notes any resistance to fungal growth by the carpet.

14.10 PLACEMENT OF ANTIMICROBIAL AGENTS

To achieve the best efficacy using AATCC Test Method 174, the antimicrobial agent must come in contact with the bacteria and fungi. If the antimicrobial agent is water soluble, it can migrate from the carpet construction to the surface of the carpet and come in contact with microorganisms. If the antimicrobial agent has very low water solubility, the placement of the agent in the carpet should be at the surface where the antimicrobial agent is exposed to the microorganism. In most instances, the treatment needs to be placed on the top or bottom side of the carpet to achieve the best results in AATCC Test Method 174.

Many applications of antimicrobial agents for carpets are in the tufting primary or latex. There have been requests from the carpet industry to modify the AATCC Test Method 174 to allow the face fiber to be shaven off to allow the antimicrobial agent to be tested. The request is based on microorganisms growing at the base of the pile fiber.

There have been numerous debates over the years concerning the contribution to microorganism growth by carpet. This particular subject matter will not be debated.

14.11 SUMMARY

Bacteria and fungus can be found in virtually every environment. Most travel by being airborne and can collect in carpet. Typically, health care facilities specify carpet that has an antimicrobial treatment. Tests have shown that properly treated carpet can prevent and inhibit the growth of bacteria.

15

Color, Dyes, Dyeing, and Printing

From the buyer's point of view, color is a major factor in the selection of carpet. Color sets the mood for the room and, at the same time, influences the care and maintenance of the room. Buyer's may avoid a light color because it would "show the dirt" or select a multicolored rug because lint would blend into the face of the carpet. Therefore, this chapter focuses on color from theory to implementation.

15.1 COLOR THEORY

Color is defined as the net response of an observer to visual physical phenomena involving visible radiant energy of varying intensities over the wavelength range 400 to 700 nanometers (nm). The net color seen by the observer is dependent on integration of three factors:

- The nature of the light source.
- The light absorption properties of the object observed.
- The response of the eye to the light reflected from the object.

The relative intensities of the various wavelengths of visible light observed by the eye are translated by the mind of the observer resulting in the perception of color. In color measurement, the human eye is replaced by a photocell, which detects the light energy present at various visible wavelengths.

Visible light is a narrow band of electromagnetic radiation from 400 to 700 nm (1 nm equals 10^{-9} meters) detected by the human eye. Radiation falling below 400 nm is ultraviolet radiation, and above 700 nm is infrared radiation; both are unseen by the human eye. If pure light of a given wavelength is observed, it will have a color corresponding to that wavelength. Pure wavelengths of light are seen when white light is refracted by a prism into a "rainbow" spectrum of continuous color. Light sources such as sunlight, incandescent light, and fluorescent light are continuums of various wavelengths of light with the relative amounts of the wavelengths dependent on the overall intensity and type of light source. Sunlight at noon has very nearly the same intensity of each wavelength of light throughout the visible spectrum, whereas at dusk sunlight is of lower intensity and has greater quantities of the longer, red wavelengths than of shorter, blue wavelengths. Fluorescent lights generally contain large amounts of shorter, blue wavelengths, while incandescent tungsten lights contain a large component of longer, red wavelengths compared to noon sunlight. Differences in intensity and wavelength distribution between light sources has a profound effect on the color observed for a carpet, since the carpet can absorb and reflect only that light available to it from the source. When a dyed carpet appears different in color or shade under two different light sources, the phenomenon is referred to as *flare*. When two carpets dyed with different dyes or dye combinations match under one light source, but not under another, the effect is called *metamerism*.

When light from a source strikes a dyed carpet surface, different portions of the light of the various wavelengths are absorbed by the dye, depending of the structure and light absorption characteristics of the dye. Light not absorbed by the dye on the carpet is reflected from the surface as diffuse light, and the observer sees the colors shown in Table 15.1. The color seen is a composite of all the wavelengths reflected from the fabric. If significant direct reflectance of light from the fabric occurs, the fabric exhibits a degree of *gloss*. If little or no light throughout the visible range is absorbed by the carpet and the majority of light is reflected, the carpet appears white. If the carpet absorbs all of the light striking it, the carpet is black. If uniform light absorption and reflectance across the visible wavelengths occur at some intermediate level, the carpet will be a shade of gray.

Table 15.1. Colors after Absorption/Reflectance

Wavelength of Light Absorbed (nm)	Light Absorbed by Dyed Textile	Color Seen by the Observer
400–435	Violet	Yellow-green
435–480	Blue	Yellow
480–490	Green-blue	Orange
490–500	Blue-green	Red
500–560	Green	Purple
560–580	Yellow-green	Violet
580–595	Yellow	Blue
595–605	Orange	Green-blue
605–700	Red	Blue-green

The dye absorbs discrete packages or quanta of light, and the dye molecule is excited to a higher energy state. This energy is normally harmlessly dissipated through increased vibration within the dye molecule as heat, and the dye is then ready to absorb another quantum of light. If the dye cannot effectively dissipate this energy, the dye will undergo chemical attack and color fading or color change will occur, or the energy will be transferred to the fiber causing chemical damage.

Organic molecules that contain unsaturated double bonds are capable of absorbing light within a given wavelength range (usually in the ultraviolet). If these double bonds are conjugated and alternate within the molecule, they are able to mutually interact with one another as a cloud of π electrons. If sufficient conjugation exists, the molecule will partially absorb light in the lower energy visible wavelength range and will be considered a *dye* or a *pigment*. In general, dyes are colored molecules soluble or dispersible in water or solvent media, which can penetrate the fiber on coloration, whereas pigments are not dispersible and must be mechanically entrapped in or locked to the fiber by a binding resin.

The unsaturated groups, which can be conjugated to make the molecule colored, are referred to as *chromophores*. Groups that enhance or alter the color within a conjugated system through alteration of the electron density are referred to as *auxochromes*. A series of typical chromophores and auxochromes can be seen in Fig. (15.1).

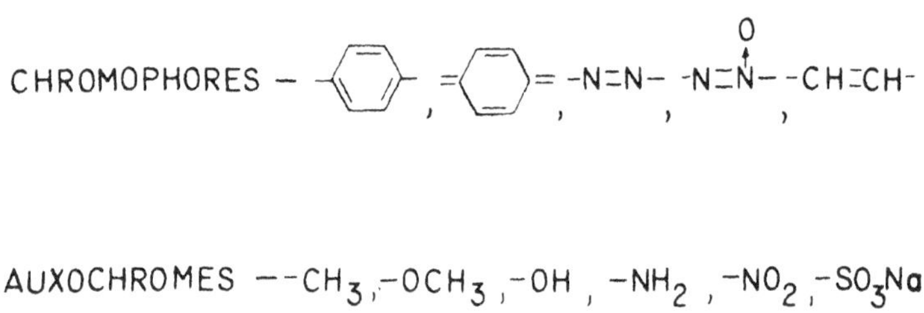

Figure 15.1. Structures of chromophores and auxochromes.

The third component in color is the observer, which can be the human eye or a photo detector in a color instrument. Most color measurement systems are based on a standard light source and a "standard observer" for quantitative measurement of color. The human eye doesn't respond uniformly to color throughout the visible region, but gives maximum response in the middle visible wavelengths. In summary, color observed is a composite of three factors:

- The light source
- The object
- The observer

It is represented in Fig. 15.2. A change in any one of these three factors will affect the net color observed.

The color of a textile can be completely defined by the reflectance color spectrum of the fabric as seen above; however, since spectra are of limited value in defining given colors or shades, numerical color systems have been developed to quantitatively or qualitatively define color. Knowing the relative position of two colors within the color space permits one to determine differences between the two colors. Color is often defined by the following terms: hue, chroma or saturation, and lightness or value. These color terms are defined as follows:

- *Hue*—basic color type (i.e., red, blue, green, etc.).
- *Chroma or saturation*—basic difference between color and the closest primary hue (i.e., difference between rose red and true red).
- *Lightness or value*—relative degree of lightness (i.e., degree of gray)

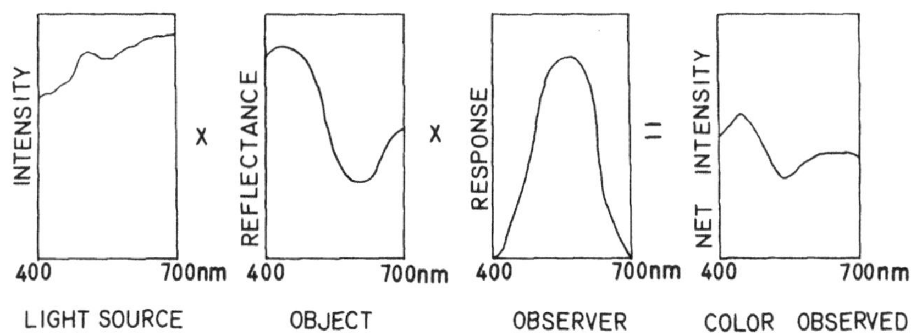

Figure 15.2. Observed color as a composite of three factors.

15.1.1 The Munsell System

The Munsell system uses standard hues and numerical values for chroma and lightness to define color. An array of chips of equal color difference as defined by many observers makes up the Munsell Book of Color. The *primaries*, defined as colors that are spectrally pure and the farthest away from neutral gray, consist of all major hues plus purple arranged logically to form a horizontal circle. The value or lightness makes up the vertical axis varying from 0 to 10 with 0 being black, 10 being white, and the intermediate numbers being varying shades of gray. The chroma is the degree of deviation from the "true" primary and can vary from 0 to 14, with the larger numbers being closest to the "pure" color. To determine the Munsell value for a given color, the Munsell color book is consulted to determine the color designation. For example, if a color had a designation of 5BG 5/8, the sample would be a blue-green with a value or lightness of 5 and a chroma of 8. In this way, color can be characterized, but it is not possible to assign strict numerical values to color differences between two dyed samples using the Munsell system.

15.1.2 Additive and Subtractive Systems

Additive color systems used in color measurement determine the amount of three color primary lights necessary to define a given color using a standard light source and observer. Dyers and colorists use subtractive systems, since the net color reflected from a dyebath or dyed fabric is more meaningful in matching dyeings and depends on the amount of the three subtractive primaries present. Color theory and color measurement are complex and are further complicated by an individual's response to physical, physiological, and psychological aspects of color.

Nevertheless, color differences can be effectively measured using additive color systems provided the light source, the observer, angle of viewing and degree of field observed is defined. In order to clarify and standardize the additive color system and color difference measurement, the Commission Internationale de l'Eclairage (CIE) was formed in 1931. The CIE has provided the definitions and standards necessary for color measurement. The primaries defined by CIE are not real colors, but are imaginary primaries used to define all colors in the color space. The amount of each of these primaries (values X, Y, Z) in a given color is used to define the shade and depth of shade for that color. Since these values are difficult to plot, the values are normalized and reduced to coefficients according to the following equations.

Eq. (15.1) $x = X/X + Y + Z$

Eq. (15.2) $y = Y/X + Y + Z$

Then the shade of a sample is defined by x and y, and the relative lightness by the Y value which is equivalent to the total reflectance of the dyed textile as observed by the human eye. Color differences (ΔE) between two samples are determined by Eq. (15.3).

Eq. (15.3) $\Delta E = (\Delta x^2 + \Delta y^2 + \Delta y^2)^{1/2}$

The color space can be mathematically presented in a number of ways and numerous color difference formulas and systems exist.

15.1.3 Commission Internationale de l'Eclairage System

Commission Internationale de l'Eclairage System (CIE) has established a color and color difference system based on rectangular color coordinates called CIELAB. This system uses L rather than Y for lightness, since L is more closely related to the response of the human eye, with black assigned 0 and white assigned 100. A and B provide the shade value of the color in rectangular coordinates with $A+$ red and $A-$ green, and with $B+$ yellow and $B-$ blue. Then, Eq. (15.4) gives the color difference.

Eq. (15.4) $\Delta E + (\Delta L^2 + \Delta A^2 + \Delta B^2)^{1/2}$

Also, cylindrical coordinates of L, C, and $H°$ may be used if desired where:

> L is lightness as defined above
>
> C is the distance from neutral gray shade ($C = A^2 + B^2$)
>
> H is the arc tangent B/A expressed on a 360° scale where $A+ = 0°$, $B+ = 90°$, $A- = 180°$, $B- = 270°$. The color is expressed in terms of hue angle (0°–360°) and the distance away from neutral gray.

The concentration of dye on carpet after a given dyeing time can be determined by three methods.

1. Measurement of the decrease in dye concentration in solution with time by ultraviolet-visible spectroscopy.
2. Determination of the dye concentration on the carpet dyed for a given time by dye extraction and ultraviolet-visible spectroscopy.
3. By measurement of the reflectance spectra of the dyed fabric followed by application of the Kubelka-Munk relationship [Eq. (15.5)].

Eq. (15.5) $$\frac{K}{S} = \frac{(1-R)^2}{2R}$$

K/S is the scattering coefficient and directly related to dye concentration, whereas R is the reflectance of the fabric at a wavelength of maximum absorption.

15.2 DYES AND DYE CLASSIFICATION

Dyes as colored unsaturated organic molecules must have affinity for fibers to be effectively applied. The dyes on fibers are physically bound to the fiber by one or more physical forces including hydrogen bonding, van der Waals or ionic forces, and in certain cases, chemically bound by covalent bonds.

Dyes may be classified in a number of ways, including color, intended use, trade name, chemical constitution, and basis of application. Of these classification methods, chemical constitution and basis of application have been most widely used. Chemical constitution indicates the major chromophores present in the dye, but does not indicate more than such structural aspects of the dye. A classification scheme for dyes has been developed and evolved for use by dyers, which is based on the method of application and to a lesser degree on the chemical constitution of the dye class. The classification scheme and major dye classes used on carpets follow:

- Dyes Containing Anionic Functional Groups
  Acid dyes
  Mordant dyes
- Dyes Containing Cationic Groups
  Basic dyes
- Special Colorant Class
  Disperse dyes

Dyes classified by this scheme are assigned standard designations according to dye class, color, and overall constitution by the Society of Dyers and Colorists in the Color Index (e.g., Acid Blue 141, Disperse Red 17).

15.2.1 Dyes Containing Anionic Functional Groups

A number of dyes, including acid, direct, mordant, and reactive dyes, contain functional groups that are sodium salts of sulfonic or carboxylic acids. These functional groups provide water solubility to the dyestuff. The dyes differ in subclassification in their affinity for fibers and/or the presence of special functional groups.

Acid Dyes. The *acid dyes* are large dyes containing one or more sulfonic or carboxylic acid salt functional groups (Fig. 15.3). These dyes are dyed onto fibers from an acid solution, since positive charge development within the fiber in acid solutions acts as a driving force for dye diffusion and migration into the fiber. Only fibers which develop a positive charge in the presence of acid, such as wool and other protein fibers, nylon, and certain modified synthetics, are readily dyed by acid dyes. Acid dyes on fibers are reasonably colorfast to light and laundering, but *mordanting* (more complete insolubilization of the dye through reaction with a metal salt) will improve the overall fastness properties of the dye. The color of the dye may be affected somewhat by mordanting; however, pre-metallized acid dyes (Fig. 15.4) are a special class of acid dyes that have been reacted with a mordant prior to dyeing and that have sufficient solubility to be dyed under conditions normally used for acid dyes.

One method of classifying acid dyes involves dividing them into three groups according to their application and wetfastness.

Leveling acid dyes. Leveling acid dyes color evenly to a moderate degree of fastness. For wool and protein fibers, they are small molecules requiring highly acidic dyebaths for good exhaustion. For nylon, leveling dyes are somewhat larger molecules and are applied from weakly acidic dyebaths.

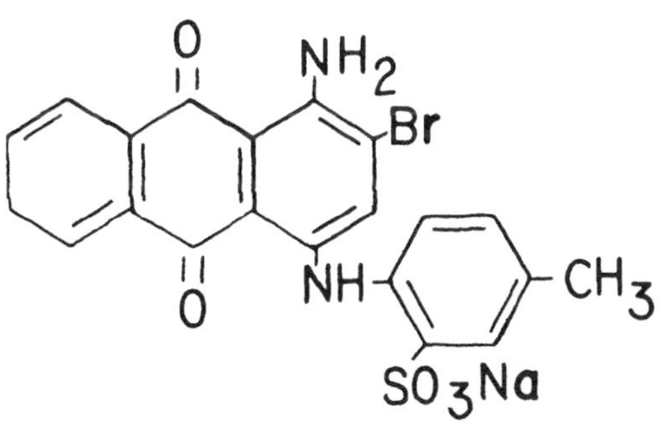

ACID BLUE 78

Figure 15.3. Structure of an acid blue dye.

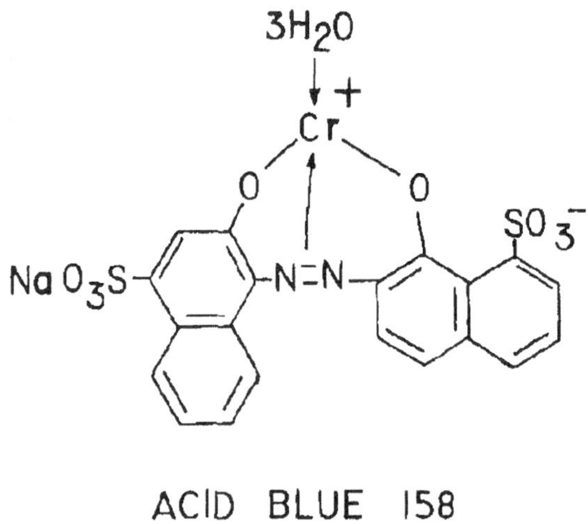

Figure 15.4. Structure of a pre-metallized dye.

Milling dyes. Milling dyes have better washfastness, but generally give duller shades and lack good leveling characteristics. They are larger than leveling dyes and are applied from dilute acetic acid solutions.

Acid milling. Acid milling or super milling dyes are applied from neutral solutions and have poor leveling characteristics due to their larger molecular size. They generally are applied to wool and have high wetfastness and lightfastness.

Mordant Dyes. *Mordant dyes* (Fig. 15.5) are acid dyes that have special sites other than acid salt anion groups that can react with a metal salt mordant. Mordant dyes are "tailor-made" to chelate with metal ions to form a strong organometallic complex of limited solubility and greater colorfastness. The fiber may be dyed initially and then mordanted (*postmordanting*), dyed and mordanted simultaneously (*comordanting*), or mordanted and then dyed (*premordanting*). Of the three methods, postmordanting is preferred. Salts of chromium, aluminum, copper, iron, tin, and cobalt are commonly used as mordants. Since the mordant affects the electron distribution and density within the dye, the color of the dyed fabric tends to change.

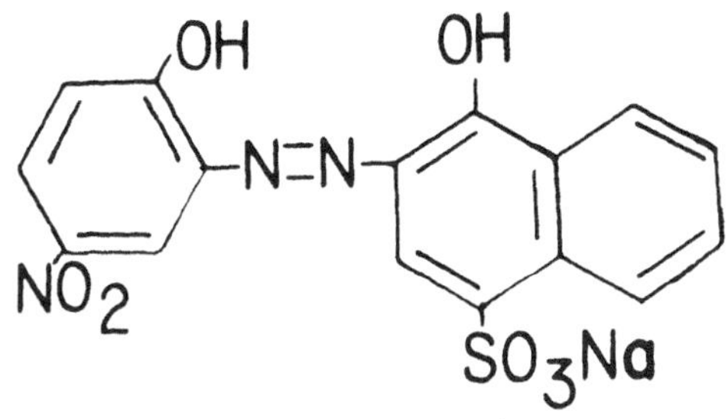

MORDANT BROWN 35

Figure 15.5. Structure of a mordant dye.

15.2.2 Dyes Containing Cationic Groups (Basic Dyes)

Basic or cationic dyes are colored cationic salts of amine derivatives. Basic dye cations will migrate toward negative charges inside the fiber (Fig. 15.6). The dyes may be applied to protein, nylon, acrylic, and specially-modified synthetic fibers. Although the dyes generally are of striking brilliance and intensity, the colorfastness of the dyes on protein and nylon fibers is generally poor. Colorfastness can be improved through mordanting with tannins, which are polyphenolic molecules extracted from wood and other natural sources, or other complexing agents. Since the insoluble parent amine is regenerated from basic dyes at basic pH, these dyes are applied from mildly acid or neutral solutions.

15.2.3 Special Colorant Classes

Several types of dyes or colorants do not "fit" logically into the other classifications and have been included in this special classification. *Disperse* dyes are small polar dye molecules which can be used to dye thermoplastic fibers such as triacetate, nylon, polyester, and other synthetic fibers.

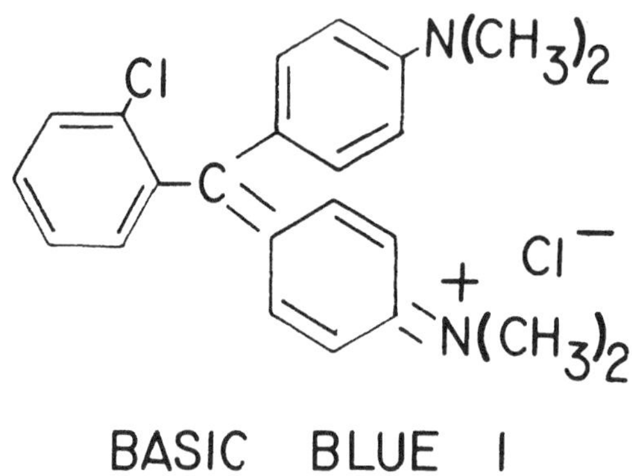

Figure 15.6. An example of a basic dye reacting with fabric.

Disperse Dyes. Disperse dyes were formulated and introduced to permit dyeing of hydrophobic thermoplastic fibers including nylon, polyester, acrylic, and other synthetics. The disperse dyes (Fig. 15.7) are small polar molecules, usually containing anthraquinone or azo groups which do not have charged cationic or anionic groups within the structure. The disperse dyes are sparingly soluble in water and must be dispersed with the aid of a surfactant in the dyebath. As the small amount of dissolved disperse dye diffuses into the fiber, additional dye dispersed in solution is dissolved until the disperse dye is nearly completely exhausted onto the fiber. The lightfastness and washfastness of these dyes is generally good, but difficulty has been encountered with fume fading from certain of the disperse dyes. Many disperse dyes have appreciable vapor pressures at elevated temperatures and can be "dyed" onto thermoplastic fibers by sublimation, which involves diffusion of the dye vapors into the fiber.

15.2.4 Dyeing of Blends

Carpet is sometimes blends of more than one fiber type. Blends containing similar fibers are easy to dye with dyes of similar structure and application characteristics. Care must be taken in dyeing blends containing fibers of highly different dye affinities, if the dyeing is carried out in the same dyebath. The dyes and their auxiliaries must be compatible with one another. When fibers in a blend are dyed the same color, the dyeing is referred to as *union dyeing*, whereas dyeing fibers in a blend of different colors is referred to as *cross dyeing*. In blends, interesting tone-on-tone, tone-on-white, and differential dyeings are possible by selection of appropriate dyes and dyeing conditions.

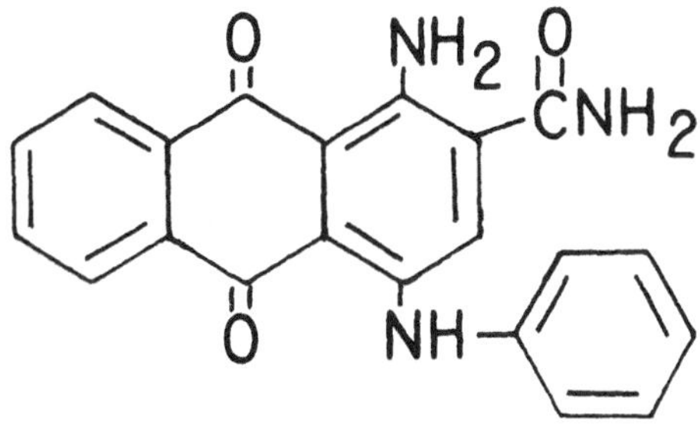

Figure 15.7. A disperse dye.

15.3 APPLICATION METHODS AND FACTORS AFFECTING DYEING

Dyes may be applied to textile structures in a number of ways and at a number of points within the textile construction process. Dyeing or printing techniques can be used. Dyeing methods involve application of dye solutions to the textile, whereas printing can be considered a specialized dyeing technique. In printing, the concentration of dye is higher, and the dye medium is thick and viscous to limit dye migration on the fabric, permitting formation of a design or pattern.

Fiber or stock dyeing involves dyeing of the loose fibers, fiber top, or sliver before yarn formation. Colorants can be added to man-made polymers before the fibers are formed to ensure uniform coloration throughout the fiber. This technique is referred to as *solution* or *dope dyeing*. Yarns or skeins can be dyed, and piece dyeing of textile fabrics can be carried out.

Discharge processes are used for application of dyes to carpet substrates in the form of fiber sliver or top, yarn, or by batch (discontinuous) or continuous methods.

15.3.1 Dyeing Methods

Fiber or stock dyeing of sliver or top is usually conducted in large vats with movement of the dye liquor through the fibers to assure intimate contact of the dye liquor with the fibers. Yarn or skein dyeing is conducted by suspending skeins of yarn in an agitated dyebath with possible additional movement of the skeins during dyeing. Yarns can also be wound as packages on perforated spindles or spools and immersed in the dyebath. The dye liquor is then circulated back and forth through the packages. Carpet rolls can also be dyed by this *package dye* technique. Package dyeing is often carried out in closed systems at elevated temperatures and pressures. *Jig dyeing* involves passage of a carpet piece back and forth from one spindle to another through a dyebath, whereas a *dye beck* containing a winch is used to move a continuous piece through the dye liquor. The above techniques, all batch processes, are graphically represented in Fig. 15.8.

Various specialized techniques have been developed for application of disperse dyes to polyester. Unless the dyeing is carried out at 100°C or above, the rate of dyeing is very slow. Dyeing with disperse dyes from aqueous solutions at 120°–130°C to achieve rapid dyeings is common, but requires the use of closed high-pressure equipment. Recently, jet dyeing has been introduced which permits not only high-temperature dyeing, but also impingement of the dye onto the moving carpet through use of a venturi jet system. Also, carriers can be introduced to permit dyeing of polyester at atmospheric pressure and below 100°C.

Carriers are usually aromatic organic compounds that can be emulsified in water and which have affinity for the polyester. The carriers penetrate the polyester, open up the molecular structure of the fiber (often resulting in swelling of the fiber), and aid in passage of the disperse dye across the dye solution-fiber interface and within the fiber. Suitable carriers include aromatic hydrocarbons such as diphenyl and methylnaphthalene, phenolics such as *o*- and *p*-phenylphenol, halogenated aromatics such as the di- and tri-chlorobenzenes, aromatic esters including methyl salicylate, butyl benzoate, and diethylphthalate, and benzaldehydes. Carriers must be removed after dyeing, and the presence of carriers in mill effluents presents a problem because of their toxicity.

Continuous dyeing is carried out on a dyeing range where carpet is continuously passed through a dye solution of sufficient length to achieve initial dye penetration. The dye on the carpet is fixed by subsequent steaming of the substrate. Foamed dye formulations have been applied to carpet, thereby effectively reducing the dye liquor to yarn ratio and reducing energy and effluent treatment costs.

A novel approach to continuously dyeing polyester with disperse dyes involves sublimation of disperse dye under heat and partial vacuum into polyester by the technique called *thermosol dyeing*. Polyester containing disperse dye applied to the fiber surface is heated near 200°C under partial vacuum for a short period of time. At this temperature, the molecular motion within the polyester is high, permitting the dye vapor to penetrate into the fiber. On cooling, the disperse dye is permanently trapped and fixed within the fiber.

15.3.2 Printing Techniques

In printing, the printing paste is applied through use of direct transfer dye using a block or engraved roller or through application using a partially marked flat or rotary screen and squeegee system.

Printing of polyester by disperse dyes can be accomplished by heat transfer printing, which is a modification of thermosol dyeing. In this process, disperse dyes are printed onto paper followed by bringing the polyester carpet and printed paper together under pressure with sufficient heating to cause diffusion of disperse dyes into the polyester. Block, flat screen, and heat transfer processes are batch processes, whereas engraved roller and rotary screen printing are continuous processes. Special techniques using dyeing solutions which give printed-style carpets have been developed. In these processes, multiple jets of different dye solutions are sprayed in programmed sequence onto the carpet as it passes under the jets to form patterns with definition very nearly like that of prints. Specialized techniques for formation of patterns on carpets have been developed. In one process, dye solution is metered and broken or cut into a pattern of drops, which are allowed to drop on a dyed carpet passing underneath to give a diffuse over-dyed pattern on the carpet. Representations of the printing and printed style processes are found in Fig. 15.9.

Figure 15.8. Dyeing processes.

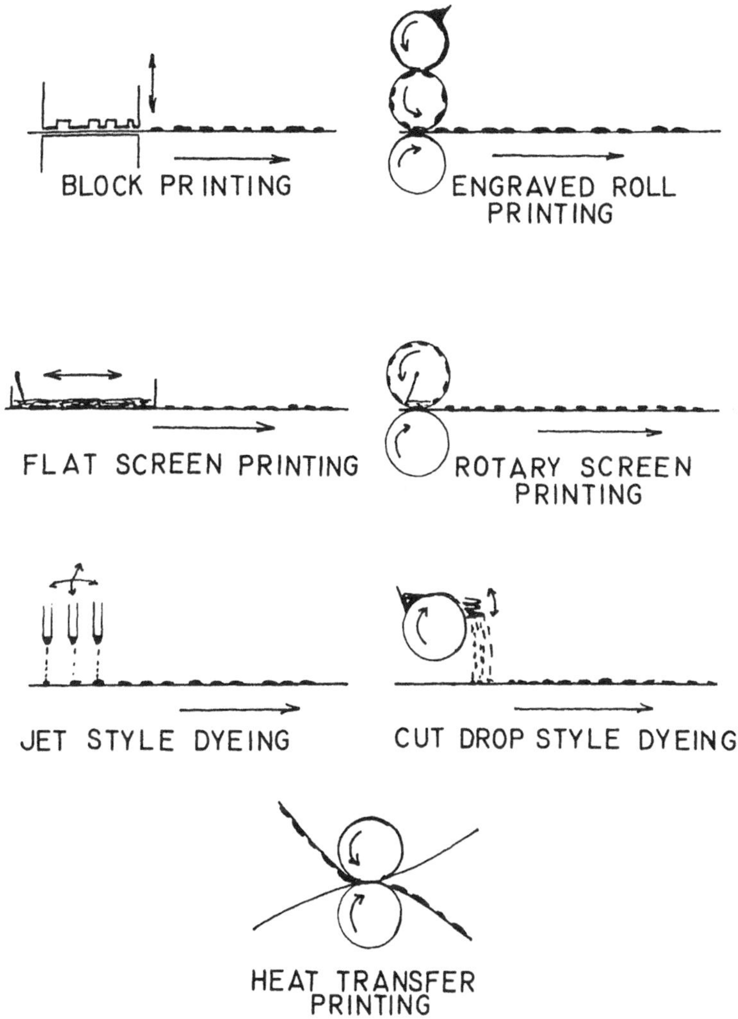

Figure 15.9. Printing and printed style processes.

15.3.3 Physical Factors Affecting Dyeing

Physical factors such as temperature, agitation, and auxiliary chemicals added to the dyebath or printing paste can alter the rate of dyeing (dyeing kinetics) and/or the total dye absorbed by the fiber (dyeing thermodynamics). In dyeing, the rate of dyeing of the fiber is dependent on the rate of migration of dye in solution to the fiber surface, the rate of diffusion of dye at the fiber interface, and the rate of diffusion of dye in the fiber matrix. Agitation of the dyebath effectively eliminates the effect of dye diffusion to the fiber in the dyebath. The rate of dye passage across the fiber-liquid interface is rapid in most cases; therefore, the rate of dyeing is solely determined by the rate of dye movement within the fiber matrix. As the temperature of dyeing is raised, the rate of "strike" of dye onto the fiber and diffusion in the fiber increases, whereas the total amount of dye present in the fiber at equilibrium decreases. In other words, heating a dyebath speeds dyeing, but decreases the total dye exhausted on the carpet. Dyeing is usually carried out at a temperature above the glass transition temperature (T_g) of the fiber, since the molecular segments of the polymers within the fiber have more mobility and permit more rapid dye diffusion above that temperature.

15.3.4 Chemical Reagents

A number of chemical reagents (auxiliaries) are added to the dyebath to alter in some manner the course of dyeing. Acids or bases may be added to the dyebath to induce charge formation in the fiber in order to increase dye diffusion. Common inorganic salts such as sodium chloride or sodium sulfate may be added to a dyebath and act as leveling agents to retard the rate of dyeing and give a more even dyeing. Surface-active agents such as anionic detergents can act as leveling agents and also permit the rapid and complete fiber wetting needed to give even dyeing. Water-softening chemicals such as sodium hexametaphosphate are added to bind or chelate hard-water calcium or magnesium ions that may interfere with dyeing. Added organic solvents and/or certain chemical reagents may increase the dye solubility and rate of penetration into the fiber. Carriers are aromatic derivatives added in disperse dyeings to open up the fiber structure of polyester and other thermoplastic fibers and to increase the rate of dye diffusion. In printing pastes, natural and synthetic gums and thickeners are necessary to provide viscosity and thickness to the dye paste and to limit dye migration across the carpet face. Heating of carpets following printing is necessary to achieve good dye penetration into the fiber, and superheated steam is preferred, since wet fixation causes a faster rate of dye migration than dry heating methods.

The fastness of the dyeing will be dependent on the location of the dye within the fiber, chemical and physical forces holding the dye to the fiber, and the stability of the dye-fiber combination to environmental factors. Dye distributed in small aggregates evenly throughout the fiber is preferred, whereas surface or ring dyeing leads to poor washfastness.

15.4 DYES APPLIED TO FIBER CLASSES

15.4.1 Dyes for Protein Fibers

Protein fibers are the most readily dyed fibers due to the numerous reactive functional groups present. They can be dyed with a wide range of dyes under acid, neutral, or slightly basic conditions.

Under acid conditions, amino groups in the protein fibers are protonated to form NH or protonated amino groups. In this form, they are able to attract dyes containing acid anions including acid and mordant dyes. Special premetallized acid dyes of sufficient solubility are used to dye protein fibers to fast colors. Protein fibers complex very readily with multivalent metal cations. Acid dyes and mordant dyes may be rendered very fast by mordanting with metal salts; chromium salts are especially effective as mordants. At neutral or slightly acid pH, protein fibers may be dyed with cationic or basic dyes; however, the fastness of the dyed fiber is poor without mordanting with tannic acid or other mordants for cationic dyes.

15.4.2 Dyes for Polyamide Fibers

The polyamides dye readily with a wide variety of dyes. Since the polyamides contain both acid carboxylic and basic amino end groups and have a reasonably high moisture regain, the fibers tend to dye like protein fibers such as wool. Since the molecular structure is somewhat more hydrophobic, more regular, and more densely packed in the polyamides than in protein fibers, they also exhibit to some degree the dyeing characteristics of other synthetic fibers such as polyesters and acrylics. Polyamides, such as nylons 6 and 6,6, can be readily dyed with dyes containing anionic groups, such as acid, metallized acid, mordant dyes, and with dyes containing cationic groups such as basic dyes. Acid dyes on nylon can be mordanted effectively for additional fastness; however, the colorfastness of basic dyes is poorer and more difficult to stabilize by mordanting. Polyamides can be readily dyed by disperse dyes at temperatures above 80°C. The biconstituent fiber of nylon and polyester can be effectively dyed by several dye types due to the nylon component but, for deep dyeings, disperse dyes are preferred. Nylon 6 and 6,6 are produced in modifications that are light, medium, or deep dyeable by acid dyes or specially dyeable by cationic dyes.

15.4.3 Dyes for Polyester Fibers

Owing to their high crystallinity and hydrophobicity, the polyester fibers are extremely difficult to dye by normal dyeing techniques unless the fiber has been modified, as in the case of modified terephthalate polyesters. A limited amount of polyester is solution dyed through incorporation of dye or pigment into the polymer melt prior to spinning. It is more common to use this technique to incorporate fluorescent brightening agents into polyester. Only smaller, relatively nonpolar dye molecules can effectively penetrate polyester; therefore, disperse dyes have been the dye class of choice for the fiber.

15.4.4 Dyes for Acrylic Fibers

The nature and distribution of acrylonitrile and comonomer or comonomers in the acrylic fibers affect the overall dyeability and the classes of dyes that may be used in dyeing these fibers. Both acrylic and modacrylic fibers can be dyed using disperse dyes, with the more hydrophobic and less crystalline modacrylic being more dyeable with this dye class. The polar cyanide groups in the acrylonitrile unit of these fibers have some affinity for acid dyes and particularly mordanted systems containing copper or chromium ions. Addition of an acid or basic comonomer such as acrylic acid or vinyl pyridine as comonomer imparts improved dyeability with basic and acid dyes, respectively, for these fibers.

15.4.5 Dyes for Polyolefin Fibers

Polyolefin fibers are hydrophobic, and the molecular chains within the fiber are tightly packed. Therefore, it is extremely difficult to dye polyolefin fibers or to increase their affinity to dyes. Colored inorganic salts or stable organometallic pigments have been added to the polymer melt prior to fiber spinning to color the fibers. Also, nonvolatile acids or bases or materials such as polyethylene oxides or metal salts have been added to the polymer prior to fiber formation to increase the affinity of the fiber for disperse, cationic, acid, or mordant dyes. Polyolefin fibers can be chemically grafted with appropriate monomers after fiber formation to improve their dyeability.

16

Stain Blockers and Fluorochemicals

Stain blocking technology in the carpet industry has been primarily focused on preventing acidic stains on nylon. Nylon can discolor because it can be penetrated by water. If a negatively charged material is in the water, it can react with positively charged areas of the nylon. Polypropylene and polyester do not have the propensity to stain because they are difficult to penetrate by water and do not have a chemical site where an acid stain can bond. Therefore, this section focuses on stain blocking chemistry for nylon.

16.1 BACKGROUND

Characteristics of fiber performance differ in many ways. The resistance to soiling and staining of fibers is one of them. Similar performance is not always true and a carpet that is soil resistant may not be stain resistant. The resistance to staining and soiling does depend on the source of the contaminant. Stains generally occur in spots while soiling is generally spread over an area where there has been foot traffic. Stains generally cannot be removed with conventional carpet cleaning techniques. Carpet stains caused by liquids are determined by the relationship of the fiber's absorption of the liquid and the stain resistance of the fiber.

Comparing wool, nylon, and polypropylene, one will find that wool has a tendency to initially repel stains due to its hydrophobic nature, but it is the least stain resistant fiber. It does have very good soil resistance. Polypropylene has better stain resistance than nylon.

Topical treatments can enhance the stain resistance of the carpet, but the treatments are not durable or may not be present on the fiber. The presence of fluorochemicals offers a physical barrier to staining, but the barrier is eventually lost through time, temperature, spill height, and/or other mechanical means. Fiber producers have researched methods to provide resistance to staining from food colorings that, typically, were like acid dyes.

16.2 STAIN RESIST CHEMICALS FOR NYLON CARPET

The commercial success of stain resist technology in the carpet industry has resulted in the widespread use of complex mixtures called *syntans*, sulfonated novolacs, or sulfonated aromatic aldehyde condensation (SAC) products. In order for this type of stain resist technology to be successful, the type of mixture must be carefully selected and its characteristics tailored to the requirements of the fiber type and also the application methods and restrictions.

The main attractive forces between the stain resist and the fiber are hydrogen bonding between the uncharged polar hydroxyl groups of the stain resist and the amide linkages in the nylon and the electrostatic attraction between the sulfonic groups in the stain resist and the protonated amine end groups of the fiber. The stain resist treatment performs best when applied at a pH of less than 2.5. This is believed to be related to both increased exhaustion of the material and also the high electrostatic attraction of the nylon that inhibits diffusion into the fiber. The addition of divalent salts, such as magnesium sulfate (Epsom salt), to the treatment may often enhance the stain resist properties of the carpet. The effective concentration generally varies depending on the morphology of the substrate being treated and the concentration of the SAC in the application mixture. While the effect in not completely understood, a possible explanation is that the salt modifies the SAC micelle formation in water and increases the sorption of the SAC at the fiber/mixture interface.

Although the processes are similar, the functional objectives of dyeing and stain resist treatment are contradictory. Pretreatment of a fiber with a SAC has been shown to reduce the uptake of monosulfonated acid dyes. This class of dyes is commonly used for nylon carpets. The presence of divalent salts in the dye bath may also form insoluble precipitates with certain dyes and other anionic dyeing assistants. Therefore, the application of the stain resist is generally not compatible with the dyeing process, although it has been attempted and practiced on a commercial scale. The preferred method for application of the stain resist chemistry is by an "after treatment," after the carpet is already dyed. This minimizes any influence on the dyeing process, makes the application more easily controlled, and yields the most repeatable carpet performance.

The after treatment may be either a bath or continuous process. The most commercially significant after treatment process involves continuous application of the stain resist using a specially designed applicator, such as the Kuster Flexnip™ or Otting Thermal Chem., which is followed by a dwell period at elevated temperature using a short vertical steamer.

16.3 TECHNOLOGY AND CHEMISTRY

A fiber has many small open parts within its molecular makeup. Each one of these spaces that can be penetrated by liquid and, subsequently, a colored substance can become a stain inside the polymer.

After dyeing, the fiber has dyes that are located within the molecular structure of the polymer. However, there are still numerous areas in which a colored particle can stain the fiber. The remaining dye sites are the key to staining. The liquids spilled on carpet, such as Kool-Aid®, act as a form of redyeing of the carpet yarn. The vacant dye sites become dyed with the liquid stain.

It was once thought that staining was actually an over dyeing of the existing color of the carpet. It is actually the color being adsorbed into the undyed areas. In the mid-1970s, 3M introduced Scotchgard® as a method of providing a protective coating on the carpet to prevent stains. The stain is kept from the fiber by preventing the liquid from penetrating the surface. DuPont later introduced Teflon®. Both are fluorocarbon-based chemicals, and they act as protection against soil and stains. These products are not perfect. The fluorocarbon can be penetrated by hot liquids and/or liquids left for long periods of time. The effectiveness can be reduced due to foot traffic or repeated cleanings. If a liquid penetrates the treated fiber, it is susceptible to staining.

The stain blocker acts as a dye and occupies the undyed dye sites. Colored substances are prevented from attaching to the dye sites that were not dyed during the dyeing process. However, the stain blockers do not act to prevent liquids from penetrating the fiber like fluorocarbons. In most stain blocking treatments, fluorocarbons are included to prevent any initial wetting of the fiber by allowing the liquid to be removed quickly. By doing so, the chances of staining are reduced.

16.4 FLUOROCHEMICALS

Fluorochemicals, with their hydrocarbon backbone and fluorocarbon ends, protect carpet fibers from foreign materials such as water and oil. The fluorocarbon does this by aligning itself with the outer surface of the fiber. Fluorochemicals are available to the carpet industry in water/surfactant dispensers.

Early fluorochemicals were not compatible with stain resistant chemicals. The fluorochemicals were cationic and the stain resistance chemicals anionic. These problems have been overcome.

Typical fluorochemicals contain a perfluoroalkyl radical having 3–20 carbons and are produced by condensation of a fluorinated alcohol or fluorinated primary amine with a suitable anhydride or isocyanate. The 3M company uses electrochemical fluorination while DuPont uses telomerization to produce their fluorochemicals.

The fluorochemicals work by lowering the surface energy of the fiber and slowing the penetration of liquids. Surface energy of nylon is reduced by about 50% with the application of a fluorocarbon. Normal nylon has a surface energy of 43 dynes/cm. The lower surface energy prevents the fiber from wetting out. In order for the fluorocarbon to perform, the perfluoroalkyl groups must be oriented outward from the fiber to produce a low cohesive force/low surface energy barrier. This orientation is usually achieved in carpet mills by heating the carpet to approximately 275°F, which is the common cure temperature for the latex backing.

Fluorochemical research has shown that the optimum perfluoroalkyl groups for oil repellency are 10–12 carbon atoms long. This is not the only variable that affects repellency. The nonfluorinated segment of the molecule, the orientation of the fluorocarbon tail, the distribution and amount of fluorocarbon on the treated surface, and the geometry of the carpet also influence repellency. The fallacy exists that good oil repellency predicts good particulate soil resistance; this simply is not true. Fluorochemicals are typically applied using several methods, which include applying the material by spray or foam at the carpet mill or by having the fiber producer apply it before shipping it to the carpet mill.

16.5 STAIN RESISTANCE TECHNOLOGY

In the latter part of the 1980s, polymethacrylic acid began to appear in the patent literature as a stain blocker by itself or as a blend or copolymer with the SACs. Poly(acrylic acid) and poly(methacrylic acid) are commonly used in textile applications, such as sizing. Like the sulfonated novolacs, these materials are water-soluble. They form hard brittle films.

A 1991 DuPont patent refers to a polystyrene and maleic anhydride copolymer as having stain resistant properties for nylon 6,6 textile substrates. The polystyrene/maleic anhydride copolymer is novel due to the absence of sulfonated groups. In addition, this stain resistant material reportedly does not yellow upon exposure to UV light.

Phenol formaldehyde condensate polymers were the first stain blocker offered on the market, but there were inherent problems such as yellowing and durability. Modifications to this condensation method began to contain sulfonated dihydroxydiphenylsulfone and phenylsulphonic acid, where at least 40% of the repeating units contain an $-SO_3X$ radical and at least 40% of the repeating units are dihydroxydiphenylsulfone.

During the 1990s, Peach State Labs of Rome, GA, received several patents involving stain resistant technology. Their patents #4,940,757, #5,015,259, #5,061,763, #5,212,272, #5,223,340, and #5,310,828 have very detailed information concerning old stain resistant chemicals and their patented technology. The information from those patents is used in the following information. Sulfonated hydroxyaromatic formaldehyde condensation products marketed and sold as stain resist chemicals include Erinol NW® from Ciba-Geigy, Intratex N® from Crompton and Knowles, Mesitol NBS® from Mobay, FX-369® from 3M, and CB-130® from Grifftex Chemicals. These products did reduce the staining from some acidic colorants, but did not resist staining from coffee or products containing turmeric, like mustard. These stain resistant chemicals would over time react with UV light or nitrous oxide and turn the fiber yellow. The yellowing would be severe enough on carpet that light colors would sometimes not get treated with the stain resistant treatment.

Research continued to reduce the yellowing phenomenon. Applying the chemicals at a pH of 1.5–2.5 reduced the yellowing. Another finding was that material treated with etherified or acrylated formaldehyde phenol condensation polymers containing 10%–25% SO_3 groups and 75%–90% SO_2 groups had improved resistance to staining and yellowing.

Polymerizing an alpha-substituted acrylic acid in the presence of a sulfonated aromatic formaldehyde polymer is the newest stain blocker chemistry. Examples are 4,4'-dihydroxydiphenylsuphone and phenyl 4-sulfonic acid. Suitable substitutes include sulfonated derivatives of naphthol, naphthalene, and vinyl aromatics.

Typically, an aromatic hydroxyl compound is first sulfonated with sulfuric acid. In general, phenol or naphthol is used. The ortho and para positions of phenol are substituted with the predominant isomer 4-sulfonic. 1-naphthol is sulfonated primarily in the 4-position, 2-naphthol is sulfonated in the 2-position and 4,4'-dihydroxydiphenylsulfone in the 3'-position. Under acidic or basic conditions, the compounds are polymerized with formaldehyde. Mixtures of the sulfonated compounds can be polymerized in this manner.

In an acid, one mole of the compound is reacted with 0.3 to 0.5 moles of formaldehyde; in a base, one mole of the compound is reacted with 0.9–1.5 mole of formaldehyde. The reaction in the base leaves the polymer more water-soluble because there are more CH_2OH terminal groups on the finished product. Cross linking of the polymer chains is possible during the reaction, and the extent of cross linking is limited by steric factors and curing conditions.

Current marketed stain resistant chemicals are sold as salts. They are manufactured by reacting the sulfonated aromatic condensation polymer with a base to form a sulfonic acid salt. The finished resins are approximately 30%–40% solids in an aqueous solution, which may contain glycols. Approximately 30%–70% of the units should be sulfonated. Typically, the composition is about 50 mole % of monosulfonated aromatic units, 15 mole % disulfonated aromatic units, and 35 mole % of unsulfonated aromatic units. Sulfonated naphthalene resins have been shown to have good wear durability and impart softness to nylon fibers.

The new chemistry combines the old stain resistant chemicals with an alpha-substituted acrylic acid ($H_2C{=}CRCO_2X$), where R is a hydrocarbon, halogenated hydrocarbon, or sulfonated hydrocarbon from C_1 to C_{15}, such as phenol, naphthol, sulfonated phenol, sulfonated naphthol or a halogen, and X is hydrogen or a hydroxylated, ethoxylated, sulfonated or halogenated hydrocarbon of C_1 to C_{15}, to make the new stain resistant product. Reactions using mixtures of other alpha-substituted acrylic acids are an alternative. Combinations of alpha-substituted acrylic acid and esters of substituted acrylic acids can be made. Preparations of these are more difficult to make effective stain blockers because, if the alcohol from which the ester is made is hydrophobic, as the percentage of ester in the chemical makeup increases, water solubility and affinity for nylon are decreased. Water solubility is not affected if the alcohol from which the ester is made is hydrophilic or basic.

When the alpha position of the acrylic acid is not substituted, the chemical produced will not give stain resistance to nylon. This may be the result of the geometric formation of the poly(acrylic acid). The alpha-substituted poly(acrylic acid) is typically syndiotactic while an unsubstituted acrylic acid is typically isotactic. The syndiotactic structure allows for efficient hydrogen bonding to the nylon structure.

The ratio of mixing the sulfonated aromatic resin to the alpha-substituted acrylic acid is approximately 1 to 8, respectively. This can range from 1:1 to 1:30. Polymer initiators such as potassium persulfate, ammonium persulfate, or sodium persulfate can be used. The reaction temperature is approximately 90°C, and the reaction takes about 60 minutes to complete. Once the reaction cools, it is about 12%–15% solids in an acidic solution. Viscosity adjustments to the solution can be made with hydrotropes. Some examples are sodium xylene sulfonate, sodium cumene sulfonate, or sodium dodecyl diphenyloxide disulfonate.

16.6 APPLICATION OF STAIN RESIST CHEMICALS

Application is an important part of the effectiveness and durability of the stain resistance of the fiber. The concentration of the barrier provided by the chemical is affected by the method of application. Application procedures used today include:

- Post dye application using a Kuster Flexnip on a continuous dye range
- Single cycle in a beck or double cycle in a beck
- Yarn application for space dye yarn

16.6.1 Basic Information

Typically, each carpet mill has had to modify the technology to meet stain resistance requirements while maintaining product and styling requirements. Low pH, sufficient exposure to heat, and moisture are required to attach the chemical to the fiber; generally, a rinse and extraction process is used.

The carpet plays an important role in stain resistance performance. The polymer used in the fiber and the fiber heat history affect the affinity of the stain resistant chemical to the nylon. A more open fiber generally requires more stain blocker to protect it.

Heat setting methods used for nylon include Suessen, Superba, and Autoclave. Nylon used in a dry heat setting method like the Suessen has a less open structure than the Superba heat setting method. Nylon has the most open structure using the Autoclave heat setting method. The reason for this is dry heat setting lowers the fiber dye rate, while wet heat setting increases the fiber dye rate.

The nylon 6,6 polymer does not have the open structure of the nylon 6 polymer. In general, nylon 6,6 has a lower dye rate than nylon 6.

16.6.2 Development of the Foam Application

Airless spray application systems have been used to apply fluorochemicals to carpet for the past twenty-five years, and they have been used to apply stain resist chemicals for the past ten years. The application of the two products used to be made separately. Stain resist chemicals needed high moisture treatments in order to be sufficiently exhausted to the nylon. Becks and continuous dye ranges were the main application areas for the stain blocker. For solution-dyed carpet, there was no practical method of applying the stain blocking chemistry and the fluorochemical during the latex coating operation.

The spray application of fluorocarbons is a practical method of application because the rate of delivery is independent of the carpet production and the inexpensive equipment used to apply the liquid. Simple delivery control and a less-fine particle spray are easy to install and typically do not need any modifications. These along with the ability to use plant air to run the unit are the advantages of the airless technique over the air-atomizing unit. The system is far from perfect and is still messy.

Exhaustible process systems were set up to be compatible with either piece or continuous dyeing along with the stain resistant technology. The first generation of stain blockers required that the carpet be thoroughly rinsed after the stain resistant application to minimize any color problems. The early technology and application of stain resist chemistry and the fluorocarbons used on a conventional dye range had some undesirable problems.

- The process had to be manually controlled.
- There were large mix tank and batching requirements.
- Equipment was needed for misting of the fluorocarbon treatment.
- The stain resist system required a low pH.
- A magnesium salt was used.
- The process involved extra energy for steaming the carpet.
- Water usage increased.

In spite of these obstacles, the treatments were a success. The early stain resistant chemicals attracted soil and, therefore, fiber-branded carpet styles needed a fluorochemical be applied to the carpet.

The Kuster Flexnip applicator was the first improvement that evolved for applying the stain resistant chemicals. This system allowed the chemicals to be mixed in-line, and the system could be automated. Changes to the application rate could be made quickly for style changes. However, the process did not reduce the energy and water consumption.

Fluorochemical treatments have to be applied to the carpet so that a majority of the application is evenly distributed in the upper portion, typically 20–50%, of the pile. Increasing the density of the carpet generally reduces the amount of penetration needed and fluorochemical required. Spray applications of fluorochemicals continue to thrive because the liquid is sprayed on latex coating equipment that is typically designed with flow-through dryers, and the energy required to dry the application is kept to a minimum. The moisture add-on is about 7%–12% depending on the carpet construction. Attempts have been made in the past to correlate the amount of fluorochemical needed to yarn type and heat-set condition, but these have been unsuccessful. Fluorochemicals are applied to carpet using the carpet yarn weight, and this is the determining factor used by most carpet mills.

Stain resist chemistries are different than fluorochemicals. Stain resist chemistries are designed to cover the entire carpet pile. A correlation between the amount of stain resist chemical required based on yarn weight and pile density has not been found. The specific type of nylon and heat-set conditions account for the variability of required stain resistance needed.

Early chemistries were designed to exhaust, i.e., to be absorbed into the fiber, in a cycle similar to the dyestuff being used. Low pH and the use of a magnesium salt improved the efficiencies of these products. The stain resist chemicals were added to the dye cycle or added as an after treatment. Each treatment required heat and an after rinse.

These systems could be added to existing equipment, but there was always extra processing time, energy consumption, space limitations, and effluent increases. Availability of space determined the exact type of application used by each mill.

The next step in the development of the stain resist application was made by trying to apply the stain resist chemical and fluorochemical finishes at the same time. It was thought that this would offer improvements in manufacturing economics, plant aesthetics, water usage, and effluent. The best method to apply the chemicals was thought to be a foam applicator.

Several problems had to be overcome if this was to be successful. The foam application system was very expensive, and Flexnip applicators were already in use and efficient in their application. The stain resistant application still needed to go through a rinse cycle.

From a processing and performance standpoint, other variables arose that had to be resolved. The fluorochemical had to vary with the weight of the carpet and be constant from fiber type to fiber type, while the stain resist application needed to remain constant regardless of the carpet weight, but vary depending upon the fiber type and heat setting. The chemical systems of both had to be compatible. The fluorochemical needed to remain on the upper portion of the pile while the stain resist had to penetrate the whole pile. Either the stain resist chemistry had to be made to not require an after rinse or the fluorochemical had to be able to remain on the yarn in an after-rinse process. Generally, the fluorochemical, unless it is bonded to the fiber, will not work in the last scenario.

Because the first generation of stain blocking chemistry required rinsing, it was not until the second-generation chemicals were developed that co-application could even be considered. The ideal situation was to put the chemicals together at the place where the fluorochemicals were being applied on the latex coating lines.

Achieving total pile penetration of the stain resist chemical was difficult. Experience had shown that fluorochemicals did not penetrate the entire pile. Fluorochemicals on the market today are emulsions, but stain resist chemicals are solutions. Solutions tend to migrate better than emulsions so variable penetration of the two materials was thought possible.

Foam applications have the following advantages associated with them:

- Automated control of the application according to speed changes.
- Automated control of the add-on of the fluorochemical based on pile weight.
- Automated control of the stain resist based on the fiber type.
- Reduction in effluent, water usage, and energy.
- Proper application of the chemical components on the carpet pile.
- Improved working conditions for the mill employees.
- Automated in-line mixing.
- Simplified operator controls.

The true co-application process has never been successfully accomplished on a mass production scale. Most applications today still use a small stain resistant application first, followed by a steam treatment, and then the fluorochemical is applied by foam or spray. This generally results in a lower total chemical cost than using the stain resistant chemicals by themselves. The addition of a small amount of stain blocker by conventional methods before applying the fluorochemical is called *priming*. The most conventional applicator of the stain resist chemical today is the Flexnip.

The differences in converting an existing spray system to a foam system are more extensive than what has been described in this text. There are problems associated with using two or more chemicals together in the same system, especially in a foam application. Some chemicals do not foam well and require foaming agents. The foaming agent must be chosen based on its chemical compatibility, foaming ability, and lack of any side effects. Most foaming agents have a tendency to increase soiling dramatically. Also, the volume of air introduced into the chemical mix must be correct so the chemicals are positioned properly in the carpet pile. The key to the success of these systems lies in the computerization of the many controls needed to operate the foam unit.

16.7 EXAMPLES OF STAIN RESISTANT APPLICATIONS

16.7.1 Batch Exhaust

This application can be made before or during the dye cycle. A minimum of 0.3% solids of the stain blocker based on the weight of the nylon is added to the dye bath after the dyeing of the carpet. The pH is adjusted to 2.0 to 2.5 at the beginning of the cycle and should not go above 5.5. The exhaust time for a pH of 2 at 160°F is approximately 15 minutes. The carpet is rinsed and then dried.

Suitable acids for stain blockers are sulfamic, formic, or citric.

16.7.2 Continuous Exhaust

The stain blocker is applied to the carpet using a Flexnip applicator at a 300% wet pick up of a 1% stain blocker solids solution. A minimum of 0.3% solids of the stain blocker based on the weight of the nylon is recommended. The pH of the stain blocker is adjusted to 2.0 to 2.5 prior to the application. The carpet is steamed for 1–2 minutes to fix the stain blocker and then rinsed and dried.

Alternative methods of application are spray, foam, or printing.

16.7.3 Use in Drying Only

This method is similar to continuous exhaust except the carpet is not processed through a steamer and is dried only.

16.7.4 Foam Application

Foam generation is normally achieved by adding a foaming agent into a solution containing between 1:1 and 1:10 of stain blocker to water. A typical foaming agent is ammonium laurel sulfate.

16.7.5 Exhaustion of Stain Blocker on Nylon

Research conducted by the Textile Research Institute on the exhaustion of a stain blocker on nylon and its effect on stain-blocking performance is reviewed below. Although nylon had good stain-blocking effects against acid stains at low exhaustion levels, stain blocking against coffee stains is improved with increased exhaustion. For instance, a stain blocker with as little as 25% penetration into yarn can achieve an excellent stain-blocking rating against Kool-Aid, but have almost no stain-blocking effect against coffee. Some stain blockers require up to 50% penetration into the fiber to have good stain resistance against coffee.

The stain blocker itself along with pH, time, and temperature controls the exhaustion of sulfonated aromatic formaldehyde condensate stain blockers on nylon. To illustrate this, a test using 10 grams of nylon 6,6 yarn was immersed into a water solution containing 0.5 grams of Mesitol NBS at various times, temperatures, and pH. The pH of the liquid solution was adjusted with sulfamic acid. The penetration of the stain blocker was examined across the cross section of the fiber after testing. The results are shown inTable 16.1. The data show the biggest influence of penetration into nylon by the stain blocker is pH, followed by time, then temperature.

Other factors that contribute to the penetration of stain blockers into nylon fibers are molecular structure of the stain blocker, molecular size of the stain blocker, and the yarn pretreatment prior to the application of the stain blocker. If the molecular structure of the stain blocker is the same, molecular size and yarn pretreatment are the main factors that affect yarn penetration. Autoclaved and Superba heat-set yarn allow further penetration of the stain blocker into the yarn than Suessen heat-set yarn. The wet heat-setting methods open the fiber and allow the stain blocker to penetrate the nylon. The differences in penetration of the stain blocker into wet and dry heat-set fiber is reduced as the molecular size of the stain blocker is increased.

Table 16.1. Test Results on Stain Blocker Exhaustion Using Time, Temperature, and pH

Test Number	Temp °C	Time (minutes)	pH	% Penetration
1	80	60	3	56
2	100	20	3	54
3	80	20	2	53
4	100	20	6	49
5	80	20	3	47
6	80	20	3	46
7	80	20	3	46
8	80	20	4	43
9	80	60	6	41
10	80	20	5	41
11	80	10	3	35
12	80	5	3	35
13	80	10	6	34
14	80	20	6	34
15	50	20	3	29
16	80	5	6	28
17	50	20	6	28
18	80	20	7	25

16.8 PERFORMANCE TESTING

Several factors working in combination with each other affect the performance of stain resistant applications. They are the stain resistant chemical type and amount, the application procedure placing the chemical on the fiber and the barrier formed, the nylon polymer type and its heat history, and anionic fluorochemicals.

The performance of various carpets on the market can vary greatly because of these factors. In general, there is an improvement in stain resistance of the carpet versus untreated carpet.

In the carpet industry, the new test standard to measure stain resistance is AATCC 175-1991. This test method uses liquid containing FD&C Red Dye #40. The test involves putting the food coloring on the carpet, letting it stand for twenty-four hours, then washing it out. The residual stain, if any, is visually rated on a numerical scale from 10 = no staining to 1 = severe staining. A rating of 8 to 9 or greater is normally considered acceptable.

Part 5

Performance, Cleaning, and Recycling

17

Performance Issues

The ultimate test of the success or failure of the carpet rests in its performance. This chapter focuses on the various measures to test performance and predict performance before the carpet is installed, flammability treatments that are required by law, the ability of the carpet to absorb sound, and a multitude of other factors that affect the performance of the carpet.

17.1 STANDARDS AND TESTS

Standards and tests for carpet and its backing are set by the U.S. Department of Housing and Urban Development (HUD). Test methods are published by the American Association of Textile Chemists and Colorists (AATCC), and the American Society for Testing Materials (ASTM). Carpet performance based on test results can be obtained by using various laboratories and test methods. A specific list of tests can be found in the Carpet Test Method section at the end of the book.

17.1.1 Standards and Tests for Carpets

Tuftbind. The U.S. Department of Housing and Urban Development (HUD) sets the minimum tuftbind for carpet. This falls under HUD UM44D. The minimum requirement is 6.25 pounds for loop carpet and 3 pounds for cut pile.

Lightfastness. To test the lightfastness of yarn, most carpet manufacturers and their customers specify AATCC Test Method 16 E.

Crockfastness. To test the colorfastness to abrasion of a fabric, most carpet manufacturers and their customers specify AATCC Test Method 108.

Atmospheric Fading. The American Textile Chemists and Colorists Test Method 23 tests carpet for fastness to atmospheric contaminants while AATCC Test Method 129 tests for ozone fading.

17.1.2 Standards and Tests for Backing

Latex Coating. Tuftlock or tuftbind is used to measure how well the latex is holding the yarn to the carpet. The test method is ASTM D1335. The minimum requirement is 6.25 pounds for loop and 3 pounds for cut pile. At one time, the minimum requirement was 12 pounds for loop and 5 pounds for cut pile.

Peel or Delamination Strength. This is a measure of how much force is required to separate the secondary backing from the carpet. The test method used is ASTM D3936.

Tear Strength. Sometimes requests are made to determine the tear strength of carpet. ASTM D2261 is used to measure tear resistance.

17.2 APPEARANCE RETENTION

This section introduces a rough guide to predicting the appearance retention of nylon carpet. It is a modified version of a Dupont appearance retention method. It should not be used as the sole predictor of carpet performance due to the other variables that impact carpet appearance. Five factors are considered in this simple prediction method. These are face fiber, density, texture, design, and color.

To use the prediction guide, a number in Table 17.1 that corresponds to the description of each of the five factors is selected. The total of these numbers is found and divided by ten. The number that is calculated is compared to the end-use rating at the bottom of the prediction table. This gives an indication of the appearance of the carpet that will be retained.

The sum of the values selected divided by ten reveals the predicted ability of the carpet to retain its "just bought" appearance when compared to the end uses in Table 17.2.

The fact that fibers made from similar polymers are not the same needs to be kept in mind. Besides the shape, yarn denier, filament size, and other characteristics, the method used to produce the fiber contributes to performance. For example, when comparing nylons produced in one-step and two-step methods, the single-step method does not produce the same crystalline structure as the two-step process. The one-step processed nylon will be more prone to abrade and stain.

It is virtually impossible to predict the retention of the new carpet appearance for every installation. Besides multiple tufting patterns, colors, manufacturing variables, maintenance of the carpet, environment, and so forth, the underlayment also affects performance. In general, any cushion does a better job of maintaining the appearance of carpet compared to the same carpet without a cushion.

17.3 FLAMMABILITY

To illustrate how materials used in carpet can affect flammability, the two test methods this chapter focuses on are the Pill Test and the Radiant Panel Test. The Pill Test simulates a lit cigarette being dropped on the face of the carpet. All carpets produced in the United States must pass this test. The test results in the radiant panel test are used to determine where a carpet can be installed when it is in an area for public use.

The International Wool Secretariat conducted flammability tests on carpet during the 1980s. Much of their evaluation is included in the results and discussion in this chapter.

Table 17.1. Appearance Retention Predictor Guide

Factor	Description	Number
Face Fiber	Name Brand Nylon (DuPont, Honeywell)	20
	Unbranded Nylon	10
Density	5,000 and up	20
	4,000–5,000	16
	3,000–4,000	12
	2,000–3,000	8
	1,000–2,000	4
Texture	Loop Filament	10
	Cut/Loop Filament	10
	Cut Pile with 5 twists per inch	10
	Cut Pile with 4–5 twists per inch	8
	Cut Pile with 3–4 twists per inch	6
	Cut Pile with 2–3 twist per inch	4
Design	Random	20
	Geometric	16
	Tweed	12
	Heather	8
	Solid	4
Color	Medium	30
	Dark	20
	Light	10

Table 17.2. Carpet Wear Predictor

End Use	Numerical Value
Heavy	7.5–10
Medium	5–7.5
Light	2.5–5

17.3.1 Pill Test

Federal Regulation DOC FF 1-70, the Pill Test, mandates carpets pass its flammability specification. The test method is described in ASTM D2859. It is referred to as the Pill Test because a methenamine tablet is used as a fuel source.

Since nylon is the most commonly used fiber, tests were conducted using the raw materials in Table 17.3. Carpet was tufted using an eight gauge-tufting machine producing a level loop pattern. The pile height (PH) of the tuft and the stitches per inch (SPI) were changed to determine their effects on flammability.

Table 17.3. Carpet Flammability Test

Raw Materials	
Face	Nylon 6
Primary Backing	Woven Polypropylene (PP)
	Nonwoven Polyester (PET)
Secondary Backing	Polyurethane Foam
	Woven Polypropylene
Latex	Carboxylated SBR latex containing 300-phr calcium carbonate
	Carboxylated SBR latex containing 200-phr aluminum trihydrate and 100-phr calcium carbonate

Latex was applied to a 9 by 9 inch carpet sample to obtain a dry coating weight of 25 ounces per square yard. All samples were dried at 350°F for 15 minutes. The carpet was placed in an open-ended 12 by 12 by 12 inch chamber. A methenamine tablet was placed in the middle of the sample and ignited.

Eight samples per construction were tested. The samples either passed or failed. Variables evaluated were pile height, stitches per inch, primary backing, secondary backing, and latex compound. The time to flame extinction and the dimension of the burn were also recorded, and the results were averaged. Tables 17.4–17.6 summarize the results.

17.3.2 Discussion of Pill Test Results

As can be seen by the data presented in Tables 17.4–17.6, as the pile height increased so did the time to flame extinction and burn length. This is most likely related to more yarn being available to act as fuel for the fire. The largest number of failures was samples made with the polypropylene tufting primary backing, polyurethane secondary backing, and no aluminum trihydrate (ATH) in the latex. There did not appear to be a correlation between stitches per inch and flammability.

The effects of the primary backing and secondary backing were evaluated together. Carpet samples made with the polypropylene primary backing had longer burn times than the nonwoven polyester. The same could be said comparing the polyurethane foam secondary backing to the woven polypropylene.

This evaluation shows carpet made with a woven polypropylene primary backing, polyurethane foam secondary backing, and no ATH in the latex results in a large number of failures for the Pill Test. The use of a nonwoven polyester primary backing compared to the woven polypropylene had better resistance to burning. This is probably related to the nonwoven polyester not melting completely as did the polypropylene. The nonwoven polyester primary backing protected the latex from burning.

The burning of the polyurethane foam was very noticeable. A bright orange flame could be seen on the burning samples. It is suspected that the porous nature of the polyurethane foam allowed air into the burning area to support combustion.

Table 17.4. Average and Rounded-off Burn Length in Inches of Passing Samples

			PP Tufting Primary Pile Height (in inches)			PET Tufting Primary Pile Height (in inches)		
SPI	ATH	Secondary	.25	.35	.45	.25	.35	.45
6	None	PU Foam	2.0	2.25	2.5	2.0	2.25	2.25
7	None	PU Foam	2.0	2.25	2.5	2.0	2.25	2.25
8	None	PU Foam	2.25	2.25	2.5	2.0	2.25	2.25
6	Yes	PU Foam	2.0	2.0	2.25	2.0	2.0	2.25
7	Yes	PU Foam	2.0	2.0	2.25	2.0	2.0	2.25
8	Yes	PU Foam	2.0	2.0	2.0	2.0	2.0	2.25
6	None	Woven PP	2.0	2.0	2.5	2.0	2.0	2.25
7	None	Woven PP	2.0	2.0	2.5	2.0	2.0	2.25
8	None	Woven PP	2.0	2.0	2.5	2.0	2.0	2.25
6	Yes	Woven PP	1.75	2.0	2.25	2.0	2.0	2.25
7	Yes	Woven PP	1.75	2.0	2.25	2.0	2.0	2.25
8	Yes	Woven PP	2.0	2.0	2.25	2.0	2.0	2.25

Table 17.5. Total Number of Test Failures

			PP Tufting Primary Pile Height (in inches)			PET Tufting Primary Pile Height (in inches)		
SPI	**ATH**	**Secondary**	**.25**	**.35**	**.45**	**.25**	**.35**	**.45**
6	None	PU Foam	0	2	4	0	0	2
7	None	PU Foam	1	2	5	0	0	2
8	None	PU Foam	1	3	6	0	0	2
6	Yes	PU Foam	0	0	0	0	0	0
7	Yes	PU Foam	0	0	0	0	0	0
8	Yes	PU Foam	0	0	0	0	0	0
6	No	Woven PP	0	0	0	0	0	0
7	No	Woven PP	0	0	0	0	0	0
8	No	Woven PP	0	0	0	0	0	0
6	Yes	Woven PP	0	0	0	0	0	0
7	Yes	Woven PP	0	0	0	0	0	0
8	Yes	Woven PP	0	0	0	0	0	0

Table 17.6. Average and Rounded Off Time in Seconds to Flame Extinction

			PP Tufting Primary Pile Height (in inches)			PET Tufting Primary Pile Height (in inches)		
SPI	**ATH**	**Secondary**	**.25**	**.35**	**.45**	**.25**	**.35**	**.45**
6	None	PU Foam	170	170	210	120	140	140
7	None	PU Foam	180	200	220	130	150	150
8	None	PU Foam	210	260	280	140	160	160
6	Yes	PU Foam	110	115	130	100	100	100
7	Yes	PU Foam	115	130	130	105	105	110
8	Yes	PU Foam	110	130	130	110	115	120
6	No	Woven PP	130	140	150	100	110	110
7	No	Woven PP	135	145	150	100	105	120
8	No	Woven PP	140	150	170	100	105	120
6	Yes	Woven PP	100	115	120	100	100	105
7	Yes	Woven PP	110	115	120	100	100	110
8	Yes	Woven PP	115	115	120	105	110	115

17.3.3 Radiant Panel Test

Although carpets may pass the Pill Test, they may not do well on the Radiant Panel Test. In this test, a radiant heat source is used to determine the amount of thermal energy needed to support combustion on carpet.

Thermoplastic fibers in carpet normally melt before a flame can ignite the pile of the carpet. This melting process, as compared to a nonmelting carpet fiber, causes a longer ignition time at high external radiant flux. A similar phenomenon occurs with tufting primaries.

While the effects of pile density and composition play a role in the Pill Test, they do not play a significant role in the Radiant Panel Test. *Critical radiant flux*, the length along the carpet that burns before the flame is extinguished, is determined by thermal energy from the flame, which maintains combustion. External energy flux determines ignition delay time.

Nylon carpet performs better in the Radiant Panel Test than carpet with other thermoplastic fibers. Nylon carpet has a slow flame velocity and longer ignition time due to the manner in which the pile melts. The molten nylon forms a layer that is much lower than the original pile height. At the same time, the thermal conductivity of the molten nylon is higher than that of the original surface. With the nylon melting, there is conduction heat loss. Nylon 6,6 tends to form a larger molten layer than nylon 6; nylon 6,6, therefore, generally performs better in this test. The face construction of the nylon carpet has little effect on ignitability and flame spread due to the conduction heat loss. The effect of a cushion underneath nylon carpet is significant in terms of ignitability and flame spread due to the change in conduction heat loss.

In the Radiant Panel Test, rapid flame spread can occur because the high temperature at the ignition point is approximately the pyrolysis temperature of the carpet. The flame spread is dependent on the pyrolysis temperature. Rapid flame spread is normally not seen unless there is a cushion underneath the carpet. The surface temperature of the carpet typically is the same whether or not there is a cushion present. The carpet backing temperature is lower without a cushion.

Typically, when comparing samples of the same type of cushion using the Radiant Panel Test, the density of the cushion has an effect on the results. As the density of the cushion increases, flammability decreases. This probably can be related to thermal conductivity and specific heat.

High pile density carpet has a higher heat loss than a low pile density carpet. The carpet with the high density does not burn as deep into the backing. Comparing nylon carpet to other carpets with a thermoplastic fiber, nylon fibers burn more completely while others leave a char due to incomplete combustion. The backing of the nylon carpet can completely burn due to the high heat of combustion.

Reduced flammability of carpet based on results of the radiant panel can be expected by increasing the filler loading of the latex, thereby reducing the available fuel for combustion.

Tests on Residential Carpet. Four different face fibers or blends were tufted into residential face constructions and tested for flammability using the Radiant Panel Test or ASTM E648. These were wool, wool/nylon blend, polypropylene, and nylon 6,6. Carpets were tufted into a woven polypropylene primary backing. They were backed with a calcium carbonate-filled SBR latex and woven polypropylene or polyurethane cushion secondary backing.

The carpet was placed at the base of the test chamber with a radiant panel positioned at a 30° incline to the carpet. An open flame ignited the carpet and was removed. The heat along the carpet being tested was measured from approximately 1.0 to 0.1 watts per square meter.

Carpets with wool fibers had the best results. The addition of nylon to the wool lowered the results. All nylon carpets burned through the backing. This did not take place with the wool carpets. All polypropylene samples burned the length of the carpet.

A trend was seen in the wool, wool/nylon blend, and nylon that indicated that increasing the density of the face fiber improved the results. There appeared to be no difference in whether the carpet was a loop or cut pile construction. The tabulated results are in Table 17.7.

Placing a cushion underneath the carpet did not have any effect on the wool carpet, but dramatically reduced the results of the nylon carpet. This difference is best explained in heat released during combustion of the two face fibers, nylon being higher than wool. Cushions act as insulators to the floor. This does not allow heat from the flaming carpet to be carried away by the floor and reduce the heat for combustion.

Tests on Commercial Carpet. Five different face fibers or blends were tufted into commercial face constructions and tested for flammability using the Radiant Panel Test or ASTM E648. These were wool, wool/nylon blend, polyester, polypropylene, and nylon 6,6. Carpets were tufted into a woven polypropylene primary backing. They were backed with a calcium carbonate-filled SBR latex and either a woven polypropylene or polyurethane cushion secondary backing.

Table 17.7. Radiant Panel Results for Residential Carpet

Carpet	Fiber Type	Total Carpet Wt (oz/sq yd)	Total Pile Wt (oz/sq yd)	Fiber Density	Radiant Panel Result*	With or Without Pad
Cut Pile	Wool	76	28	3214	0.70	0.85
Cut Pile	Wool	85	35	3600	0.70	0.60
Cut Pile	Wool	92	35	4731	0.70	0.75
Loop Pile	Wool	88	26	3877	0.60	0.65
Loop Pile	Wool	92	31	4065	0.75	0.70
Loop Pile	Wool	42	39	3231	0.65	0.70
Loop Pile	Wool/ Nylon	78	28	3360	0.65	0.60
Cut Pile	Wool/ Nylon	88	39	4011	0.65	0.60
Cut Pile	Wool/ Nylon	99	46	4476	0.70	0.65
Loop Pile	Wool/ Nylon	78	25	3214	0.50	0.65
Loop Pile	Wool/ Nylon	99	39	4529	0.60	0.60
Cut Pile	Poly- propylene	78	25	3840	0.10	0.10
Loop Pile	Poly- propylene	81	28	3360	0.10	0.10
Loop Pile	Poly- propylene	88	35	4065	0.10	0.10
Loop Pile	Poly- propylene	88	35	4065	0.10	0.10

*Avg. and rounded-off.

(Cont'd.)

Table 17.7. *(Cont'd.)*

Carpet	Fiber Type	Total Carpet Wt (oz/sq yd)	Total Pile Wt (oz/sq yd)	Fiber Density	Radiant Panel Result*	With or Without Pad
Cut Pile	Nylon	78	28	3600	0.50	0.10
Cut Pile	Nylon	85	35	4065	1.0	0.20
Loop Pile	Nylon	85	25	3214	0.40	0.30
Loop Pile	Nylon	92	25	4065	0.90	0.20
Loop Pile	Nylon	99	39	4529	1.0	0.40
*Avg. and rounded-off.						

The results presented in Table 17.8 show wool and wool/nylon blend carpet had the least amount of flammability compared to the 100% thermoplastic fibers. In decreasing order of performance were nylon, polyester, and polypropylene. The study also showed the effect of a cushion underneath the carpet with a thermoplastic face. Increased flammability from the foam was the cause. The reason for the increased flammability, as explained in the residential section, was the insulating effect of the carpet causing a decrease in heat sink.

17.4 SOUND ABSORPTION

Carpet performs as a floor covering and an acoustic material, absorbing sound transmission. When compared to hard floor surfaces, carpet reduces sound transmission in a room.

Table 17.8. Radiant Panel Results for Residential Carpet

Carpet	Fiber Type	Total Carpet Wt (oz/sq yd)	Total Pile Wt (oz/sq yd)	Fiber Density	Radiant Panel Result*	With or Without Pad
Loop Pile	Wool	92	28	6300	1.0	No
Loop Pile	Wool	99	28	6300	1.0	No
Loop Pile	Wool	92	42	5040	1.0	No
Loop Pile	Wool	123	46	5342	1.0	No
Cut Pile	Wool	88	32	2722	0.70	No
Cut Pile	Wool/Nylon	106	53	4892	1.0	No
Cut Pile	Wool/Nylon	88	32	3840	0.70	No
Cut Pile	Wool/Nylon	92	32	3840	0.60	No
Loop Pile	Nylon	70	18	4050	0.65	No
Cut Pile	Nylon	78	25	2903	0.50	Yes
Cut Pile	Nylon	60	18	3240	0.20	Yes
Loop Pile	Nylon	78	7	2100	0.10	Yes
Cut Pile	Polyester	81	39	2552	0.25	No
Loop Pile	Poly-propylene	95	14	4582	0.10	No
Loop Pile	Poly-propylene	63	11	2475	0.10	No

*Avg. and rounded-off.

Carpet can be selected for specific acoustical applications, if needed. Specific types of carpet are useful in different types of architectural structures. These areas include theaters, broadcast studios, and open office environments.

17.4.1 Testing Terminology

There are three types of acoustic measurements: *Noise Reduction Coefficient* (NRC), *Impact Noise Rating* (INR), and *Impact Insulation Class* (IIC).

The NRC number is the average of measurements of sound absorption coefficients from 125 Hz to 4000 Hz. It is a single number used to determine the effectiveness of sound-absorbing materials for noise control.

The INR sound levels are calculated as a single rating by measuring sound in a floor-ceiling assembly. The measurement determines the impact of noise by materials in an isolated room below a ceiling. The U.S. Department of Housing and Urban Development established the criterion as a minimum standard for multifamily dwellings. A rating of less than zero is unsatisfactory, while those above zero are rated as superior.

The IIC uses the same test procedure that determines INR. This is a single figure rating for sound insulation. The floor-ceiling test results are all positive, with the highest numbers having the best sound insulation. A simple way to determine an estimate of IIC results from INR results is to add 51 to the INR result.

17.4.2 Testing of Carpet

Carpets, with and without cushions, have different sound absorbing properties. Carpets are tested in accordance with ASTM C423 *Sound Absorption of Acoustical Materials in Reverberation Rooms* to determine if carpet can absorb sound over floors and results are reported as NRC.

The following are general factors influencing sound absorption properties.

- Carpets that are similar in construction, except for loop or cut pile, will result in cut carpet having better results.
- In a loop pile carpet, increasing yarn weight but, at the same time, not increasing pile height does little to improve results.
- In a loop pile carpet, increasing pile height but, at the same time, not increasing pile weight improves results.
- Any foam on the back of carpet will increase sound-absorbing properties.
- The more air that can pass through the carpet, the better the results.
- The more air that can pass through foam backings, the better the results on a weight-by-weight basis.
- When adhering foam to carpet, permeability must be maintained to have the sound-absorbing properties of both.

17.4.3 Impact Sound Insulation

Impact noise is used to determine a floor covering's ability to absorb noise in a room with a floor and ceiling. A machine imparts sound from a floor above a ceiling and a microphone below the ceiling measures the noise that passes through both the floor and ceiling construction. Test results are reported as IIC or INR. Transmission of the sound is measured in a series of frequency bands from 100 to 3150 Hz.

The following are general factors that influence sound absorption properties.

- Carpet by itself has better sound-absorbing properties as the pile weight increases.
- Carpet with cushion has better sound-absorbing properties than carpet without cushion.
- Different types of cushion materials can significantly change the results.

17.5 OTHER PERFORMANCE ISSUES

Several other factors can affect the performance of carpets. The manufacturing process, the fiber and materials used to make the carpet, the finishes applied, the installation, and/or the traffic on the carpet can affect them. A brief explanation of other performance issues follows.

17.5.1 Butylated Hydroxy Toluene (BHT) Yellowing

Butylated hydroxy toluene is used as a chemical preservative in plastics. It can be found in some carpet backings and pads. Yellowing occurs when the BHT reacts with nitrous oxide in the air. Usually, this can be reversed with a dilute solution of citric acid added after manufacture.

17.5.2 Discoloration from Stain Blockers

Some stain blockers will turn brown when silicone-based stain blockers are added to the carpet after manufacturing or when cleaning agents with a high pH are used to clean carpet. Following the carpet manufacturer's maintenance guidelines prevents this.

17.5.3 Pile Crush or Matting

Pile crush or matting can occur in areas where there is high foot traffic. The length of time before crushing occurs is based on the density of the pile yarn and the quality of the yarn. To prevent this, a determination of the need for an underlayment must be made, the correct carpet specification for the installation should be followed, and the correct maintenance procedures must be used.

17.5.4 Shading

Shading is the appearance of an area of carpet that seems to be optically different within the overall carpet. This can be described as *trafficking*, watermarking, pooling, pile reversal, and so forth. Shading does not appear often, and many times there is no standard explanation for it. Speculation on causes for shading includes humidity, static electricity, irregularities in the floor, and floor temperature among others.

Foot prints cause *foot marking*, and it is usually temporary. Vacuuming or steam cleaning removes it.

Trafficking is caused from heavy foot traffic in a small area. After this happens, it is difficult to have the pile raised permanently.

Shade differences can occur if carpet laid side-by-side has the pile direction running different ways. The installer needs to insure the pile direction is always the same.

Shade differences can also be seen when carpet is laid side-by-side and there is a pile density difference on the edge of one of the carpets. The change in density is sometimes caused during carpet manufacturing when the pile density is reduced on one edge of the carpet.

When direction of the tips of pile fibers is reversed during the manufacturing process, it causes *pile reversal*. Pile reversal typically occurs in cut pile carpets. It is not reversible.

17.5.5 Pilling

Pills are small balls of fiber formed during abrasion of the carpet. This typically occurs on cut pile carpets. It does not indicate a wear problem and comes from loose fibers in the pile.

17.5.6 Fuzzing

This is associated with loop pile carpets. *Fuzzing* may occur when the carpet is overused, is improperly maintained, has poor underlayment, or has poor latex penetration.

17.5.7 Delamination or Backing Separation

Delamination occurs when the carpet comes loose from the backing. This can happen as a result of heavy traffic or decomposition of the latex or adhesive joining the two.

17.5.8 Indentation

Indentations are compression areas in the carpet caused by heavy objects concentrated in a small area. Moving the object occasionally or putting something under the object to spread and reduce the pressure can resolve compression.

17.5.9 Static

Static electricity is not made, but stored in carpet. Static charges vary depending on the person, humidity, and the item touched. Increasing the humidity in a room can help reduce the buildup of static.

17.5.10 Grinning

Grinning occurs when the backing is seen through the carpet pile. The grin test is performed by bending back the corner of the carpet. If the backing can be easily seen, the carpet may not wear well or retain the "as purchased" appearance.

18

Maintenance and Cleaning

This chapter provides basic information on maintenance and cleaning. Most carpet manufacturers and sellers provide customers with additional recommendations on how to keep, maintain, and clean their carpet.

Prevention is the key to increasing the life of the carpet. Having a method to keep soil from coming in from the outside, such as a mat outside the door, extends the useful life of the carpet.

Sometimes traffic lanes are created where carpet and hardwood floor come together. Wax coming off of the floor and heavy foot traffic in the area are the cause.

18.1 CLEANING METHODS AND EQUIPMENT

Vacuum. The easiest method of removing dirt is to vacuum the carpet on a regular basis.

Steam Cleaning. The process is called steam cleaning, but the water never gets hot enough to form steam. Steam cleaning uses a heated detergent that is applied under pressure to draw the soil into the equipment. The machines may have rotary brushes to help remove soil.

Shampoo. Rotary brush shampoo equipment uses a detergent solution that is released into the brush. This forms foam that is worked into the carpet. Once the shampoo is dry, it is vacuumed and the dirt is removed with the remnants of the shampoo.

Foams. Foams are generally applied via an aerosol can or container and rubbed into the pile. Once the foam is dry, the residue is vacuumed.

Bonnet Cleaning. An absorbent pad, called a bonnet, is used on the bottom of a rotary machine and a detergent solution is sprayed onto the carpet. The bonnet is used to push the detergent into the carpet and remove the soil from it. The pad can be washed and reused.

Absorbent Compound. Dry compounds contain solvents and/ or detergents and are sprinkled onto the carpet. Some type of machine works the absorbing material into the carpet, and it is later removed by vacuuming.

Periodic Deep Cleaning. Periodic deep cleaning should be used to removed deep soil and stains. For deep cleaning, dry powders or dry foams, a wet rotary brush, and/or steam cleaning are used. Dry powders and foams have the advantage of drying quickly, but they do not clean deeply into the pile. A wet rotary brush gets deep into the pile, but the carpet must be cleaned evenly. Therefore, steam cleaning is the most common method of deep cleaning a carpet.

18.2 REMOVAL OF STAINS AND CLEANING HINTS

Carpets are not indestructible. It does not matter if they have stain protection or not, they are not protected from all stains.

If a spill is dropped on a carpet, it is best to use a white paper towel or napkin to blot it. Wiping the spot only causes the liquid to spread. Solid material needs to be scooped up without scraping or brushing the stained area.

Water should not be poured on the carpet. Over wetting a carpet can cause microbes to form in the carpet, the stain to come back, the separation of the carpet and the backing, and bleeding of the dyes on clothes or to other parts of the carpet.

18.3 CLEANING STAINS AND SPILLS

Once a spill or stain is noticed, it should be cleaned immediately. The most effective cleaning method is to clean stains with water or club soda, which helps dilute the stain. The stain should be blotted straight down with a clean, dry cloth. Rubbing the stain causes the stain to grow in size. High pH cleaners should be avoided because they can harm the carpet by themselves. The least aggressive cleansers and methods should be used first, and harsher ones tried if the stain persists.

18.3.1 Common Spill Removal

Table 18.1 is a guideline for the removal of spots; it is not all inclusive. The table shows some common spots that can be cleaned by following the listed steps.

Table 18.1. Spot and Stain Removal Guide

Stain or Spot	Step 1	Step 2	Step 3	Step 4
Alcoholic Beverages	DS	A	WV	WW
Blood	A	DS	WW	A
Candle Wax	NPR	DCF		
Chewing Gum	DCF			
Chocolate	DCF	DS	A	WV
Coffee	DS	WV	WW	A
Crayon	NPR	DCF		
Fingernail Polish	NPR	DS	WW	
Ice Cream	DCF	DS	A	WV
Ballpoint Ink	NPR	DCF	DS	A
Kool-Aid	DS	WV	WW	SRK
Latex Paint	DS	A	WW	SRK
Lipstick	NPR	DCF	DS	A
Mustard	DS	WV	WW	SRK
Rust	WV	DS	WW	SRK
Soft Drinks	DS	A	WV	WW
Urine	DS	WW	WV	A
Vomit	DS	WW	WV	A

Legend: DCF - dry cleaning fluid, NPR - nail polish remover, DS - detergent solution, WW - warm water, WV - white vinegar solution, A - ammonia solution, SRK - spot removal kit.

19

Recycling

Approximately 99% of all carpets removed from residential and contract installations each year go to a landfill. This accounts for approximately 4–8 billion pounds of waste when scrap from the manufacturing process is included. To reduce impact on landfills, carpet and carpet fiber manufacturers are attempting to divert carpet materials from landfills by recycling them into usable products.

19.1 DISCUSSION

The largest volume of carpet is made with nylon face fibers, polypropylene tufting primary backing, styrene butadiene rubber latex, calcium carbonate, and polypropylene secondary backing. Carpet face fibers account for approximately one half of the total weight of the carpet. Latex accounts for almost the other half, and it may contain 70%–80% calcium carbonate. Recycling of carpet has long been difficult because of the different polymeric components and mineral fillers contained in carpeting. An added difficulty for recycling carpet is the soil that is deposited into the carpet during its use.

During most carpet recycling processes, the carpet has to be reduced to individual components. This process requires additional energy which makes some rccycling programs uneconomical. Additional costs incurred during the recycling of carpet are the labor cost to pick up the carpet and to inspect it. This ensures that there are no scraps of wood or metal. The collected carpet must be compacted and then separated by yarn type before it can be sent to a recycling center.

19.2 USES

Processed carpet waste can be turned into cushions, incorporated into concrete, or formed into a variety of plastic composites. Some examples of plastic composites are automotive parts, lumber, and parking stops. In order to make the plastic composites, some of the mineral filler must be removed during a granulation process to make pellets for extrusion.

19.2.1 Plastic Lumber

Plastic lumber can be manufactured from carpet waste. This type of recycling has been popular because a minimal amount of plastic separation is required to produce plastic lumber.

The equipment configuration needed to produce plastic lumber includes an extruder, molding unit, part extractor, and controls. The extruder screw has to be capable of turning high revolutions per minute (RPM) to melt the plastic. A screw inside the extruder has to be designed to thoroughly mix the plastic and to prevent degradation. A melt temperature inside the extruder has to be between 300° and 400°F. The control system monitors the cooling, screw speed, and internal pressures within the extruder. Linear molds, mounted on a turret, rotate through a water bath. Once the plastic is cooled, it is ejected for removal.

Another method used to produce plastic lumber or wood is to shred the carpet and add a binder to it. The material can be extruded or pressed to form a wood panel.

19.2.2 Concrete Reinforcement

Recycled carpet fibers have been incorporated into concrete for reinforcement. The fibers can be mixed into the concrete in a standard drum mixer. No more than 2% by volume is recommended due to the possible loss of flexural strength of the cement composite over time. Portland cement concrete panels have been made with waste nylon fibers. The recycling reduces plastic and shrinkage cracking in the panels.

19.2.3 Asphalt Modification

Carpet waste fibers have been used as modifiers in asphalt mixtures. There was difficulty in this recycling approach when dense asphalt structures were made. Additional asphalt was needed to coat the fibers in the mixture.

19.2.4 Use in Soil

Carpet fibers have been evaluated as additives to soil to prevent erosion, to improve road conditions on unpaved roads, and to slow water run-off on sloped land.

19.3 RECOVERY PROCESSES

Two experimental solvent recovery processes to reclaim specific polymers from carpet have been developed, but no commercialization of these is known. One of the methods uses a single solvent to dissolve carpet into separate components of plastic and non-plastic material in an enclosed vessel. Non-plastic materials settle to the bottom of the recovery vessel. While this process does not produce a specific plastic, it can produce a heterogeneous plastic mixture.

Another method of solvent plastic recovery is to use multiple solvents. Each solvent is selected to dissolve specific plastic components. This is a multiple step process and each solvent must be used separately. Carpet is exposed to the solvent and then the carpet is extracted. The solvent is flashed off and the specific plastic is recovered. These steps continue until all the separate components are collected. This is more expensive, but it produces homogenous polymers.

There are several patents that have been issued for using the same or similar materials for the carpet face, tufting primary backing, and secondary backing. The purpose of these constructions was to be able to incorporate the carpet back into itself. These have included all nylon carpet, all polyester carpet, and all polyolefin carpet. Only two processes have been commercially used.

19.3.1 Individual Carpet Component Separation Through Super Critical Fluid Separation

Carpet components can be dissolved in carbon dioxide in a super critical phase. Once the components are dissolved, thermoplastic materials are collected and separated in various steps. The latex, typically SBR, and calcium carbonate are left as waste.

19.3.2 Compatibilization of Carpet Components

This process uses compatibilizers such as maleic anhydride-grafted polypropylene by itself or with polystyrene block co-polymers under heat to form composite plastics. Using traditional types of plastic compounding equipment, these materials can be used for extrusion, injection molding, and so forth.

19.3.3 Depolymerization

A depolymerization process, in which nylon is placed into a solution of sodium hydroxide with benzyltrimethylammonium bromide (BTEMB) used as a phase transfer catalyst, has been shown to be a possible method.

19.4 EXAMPLES OF CARPET RECYCLING

19.4.1 Carpet Used for Energy

Carpet is used as a fuel supplement for coal in industrial and utility boilers.

19.4.2 Company Specific Examples of Recycling

Hoechst Celanese commercialized Trevira One® using polyester fibers tufted into a needle-punched, nonwoven polyester primary backing. The secondary backing was nonwoven polyester with low melting fibers to lock the carpet fibers in place. This construction did not require any adhesive. There were production problems with different carpet manufacturers and production of the product stopped.

Shaw Industries introduced two carpet styles called GreenHow's® and EcoWorx®. The backing contains all polyolefin components. GreenHow's has polyolefin yarn tufted into a polyolefin tufting primary backing and extrusion coated with a polyolefin secondary backing. The carpet is claimed to be 100% recyclable into the backing. Shaw has researched and incorporated carpet waste into concrete.

United Recycling Inc. has used recycled carpet as an ingredient for manufacturing carpet tack strips.

In a process that uses carpet tile, Collins and Aikman has taken the carpet and reduced it to a small particle size. The small particles are combined with post consumer low-density polyethylene to improve the properties of extruded articles. The material is used to make lumber and parking stops. Collins and Aikman produces a carpet with Powerbond ER3® backing. The backing is made from recycled materials. These materials come from their competitors, as well as their own. The material is ground and extruded into a calendar to make a sheet. This sheet is then laminated to the back of the carpet.

Several approaches have been made to recondition carpet for resale. In 1994, Milliken began its Ennovations® program to clean, recolor, and reinstall carpet tile. The cost of the carpet program is reported to be about one half of the cost of purchasing new carpet tile.

Polyester from recycled sources including soda bottles is used by Shaw Industries to make carpet fibers. It is possible to recycle up to 50% of the carpet used in such a manufacturing process. Geotextiles fabrics are made from old nylon carpet fibers. These fabrics are used for soil stabilization and in concrete mixtures for reinforcement. Shaw Industries also produces a nylon 6 fiber called EcoSolution Q® which is claimed to have 25% recycled nylon content.

Wellman Incorporated recycles polyester from various sources including soda bottles. Some of the recycled polyester is turned into staple fiber for carpet.

Image Industries Incorporated reportedly produces approximately 160 million pounds of polyester a year from recycled plastic bottles. A majority of the recycled polyester is used in broadloom carpet.

There are several carpet recycling programs in the United States that are set up to separate and recycle nylon from carpet. One program uses recovered nylon to manufacture felts for soundproofing.

Fiber producers have set up collection centers in the United States. These collection centers take carpet to recycling facilities to separate the yarn from carpet. DuPont and Honeywell have programs such as these.

Recycling nylon has been extensively researched for several years, but technologies have been slow to be commercialized. Recently, Honeywell and DSM Chemicals North America have constructed a facility to separate nylon 6 from carpet and produce caprolactam. Caprolactam is the polymer used to make nylon 6. Heat and steam are used to separate the components of carpet. During this process, caprolactam is recovered. This process is supposed to produce 100 million pounds of caprolactam from 200 million pounds of used carpet. The fiber made from this process is called Infinity.

The other materials recovered from this separation, calcium carbonate and other polymers, are planned for use as a feedstock for a cement kiln.

BASF patents describe a method to recover caprolactam from carpet with nylon 6. The nylon 6 is depolymerized to produce caprolactam using steam and an acid catalyst. Most of the backing is sent to a boiler for incineration. Ash from this process is described as suitable for plastic parts. Accepting carpet through recycling centers, BASF called this program Six Again®. BASF SAVANT® is described as an advanced, engineered nylon for cradle-to-cradle recycling.

DuPont has a similar program for recycling carpet. Separated nylon is processed through a pilot ammonolysis facility that produces basic nylon ingredients. DuPont has a patent to depolymerize nylon 6,6 to adipic acid and hexamethylenediamine.

Monsanto has patents involving the recycling of the total carpet composition into plastic pellets. These pellets are intended for automotive and industrial applications.

Georgia Composites has a process that separates carpet components to plastic resins. These thermoplastics are used in composites.

WTe Corporation has evaluated the use of carpet waste as a supplement for coal as a fuel source in industrial and utility boilers.

Appendix: Carpet Test Methods

Table 1. Carpet Test Methods

Type of Test	Test Name	Test Method
Force Required to Pull Tuft	Tuft bind	ASTM D1335
Force Required to Separate Backing from Carpet	Delamination	ASTM 3936
Flammability	Pill Test	CPSC FF 1-70
Flammability	Radiant Panel Test	ASTM E648
Smoke Density	Smoke Generation	ASTM E662
Appearance Retention	Vetterman Drum	ASTM D5417
Appearance Retention	Hexapod Drum	ASTM D5252
Appearance Retention	Roller Caster Chair Test	DIN 54324
Appearance Retention	Philips Chair Test	N/A
Colorfastness	Lightfastness	AATCC 16E
Colorfastness	Crocking	AATCC 165
Colorfastness	Ozone	AATCC 129
Colorfastness	Nitrogen Oxide	AATCC 129

(Cont'd.)

Table 1. *(Cont'd.)*

Type of Test	Test Name	Test Method
Static Generation	Static Control	AATCC 134
Stain Resistance	Stain Resistance	AATCC 175
Soil Resistance	Fluorine Test	CRI TM-102
Volatile Organic Compound	Indoor Air Quality	ASTM D5116
Density	Density	ASTM D1505
Liquid Displacement	Density	ASTM D792
Cohesive Strength	Tensile Strength	ASTM D638
Stiffness	Tensile Modulus	ASTM D638
Hardness	Hardness	ASTM D2240
Ring & Ball Softening Point	Softening Point	ASTM E28
Viscosity of Thermoplastic	Melt Index	ASTM D1238

Glossary

A

Abrasive Wear - Damage to carpet from foot traffic or other rubbing friction.
Acid Dyeable Nylon - Nylon that can be dyed with acid dyes.
Acrylic Fiber - Fibers containing copolymers of acrylonitrile.
Adhesive Strength or Peel Strength - Measure of bond of adhesive to a surface.
Adipic Acid - Chemical used in producing nylon 6,6.
Aesthetics - Perception of carpet appearance.
Affinity - Chemical tendency of chemicals to combine.
Agar - Dried extract used as a base for microbial growth.
Air Entangling - Combining multiple end of yarn with air.
AATCC - American Association of Textile Chemists and Colorists.
ASTM - American Society for Testing and Materials.
Antimicrobial - Chemical that prevents the growth of microbes.
Antistatic - Resistance to static electrical charge.
Aprotic Solvents - Solvents that do not donate protons (H^+).
ATH - Aluminum trihydrate, a commonly used fire retardant.
Atmospheric Fading - Fading of carpet color by atmospheric contaminates.
Autoclave - Equipment used to heat set yarn under pressure with super heated steam.
Average Pile Density - Weight of yarn by volume of carpet.
Average Pile Density Formula - $D=(W \times 36)/H$, W=weight, H=height.

B

Back Coating - Applying an adhesive to the back of carpet.
Backing - Materials on the back of carpet.
Backing Systems - Types of total backing constructions.
Bacteria - From bactillaceae, rod shape microbe.
Bale - Approximately 650 pounds of staple fiber in a container.
Barrel - Section of extruder, which the extruder screw resides.
BCF - Bulk continuous filament.
Beams - Cylinders, which are large, and horizontal that hold yarn for a tufting machine.
Bearding - Long fiber fuzz on loop pile carpet.
Beck - Typically a tank and reel vessel used for dyeing carpet in rope form.
Beck Dyeing - Dyeing of greige carpet in a beck, which contains dyes.
Bleeding - Loss of color when carpet is wet.
Blend - Two or more fibers or yarns.
Blending - Mixing of staple fibers.
Branded Fibers - Yarn or fiber warranted by a manufacturer.
Breaker Plate - A disc used to support screen packs.
Brookfield Viscosity - Standard measurement of viscosity expressed in centipoise.
Bright - Opposite of dull in describing fiber luster.
Broadloom - Carpet wider than six feet.
Bulking - Crimping or texturizing yarn to increase volume.

C

Cabled Yarn - Yarn made by twisting two or more yarns.
Caprolactam - Single basic chemical used to produce Nylon 6.
Calendar - A machine equipped with three or more heavy internally-heated or -cooled rolls revolving in opposite directions, that is used for continuously sheeting or plying up rubber compound, frictioning, or coating fabric with rubber compound.
Carding - Blending staple fibers by combing.
CRI - Carpet and Rug Institute.
Catalyst - Chemical that increase the reaction of polyurethane.
Cationic Dyeable Nylon - Nylon which dyes with cationic dyes.
Cohesive Strength - Internal strength of polymer, ability to resist splitting within itself.

Compatibility - Ability of mixture or compound to stay homogeneous.
Cleanability - Ability to remove soil or stain from carpet.
Coating - Application of latex or other material to the back of carpet. Typically used to hold tufts of yarn into the tufting primary.
Color Matching - Coordination of color and shade.
Colorfastness - Ability of carpet to retain color.
Construction - Specific detail of carpet manufacture.
Continuous Dyeing - Dyeing of carpet while it processes through a continuous dye range.
Continuous Filament - Unbroken strand of fiber.
Continuous Heat Setting - Process of applying heat to set bulk or twist to yarn.
Cotton Count - Yarn numbering system based on length and weight for staple fibers.
Creel - Rack or frame behind a tufting machine, which holds yarn.
Crimp - Non-linear configuration in yarn.
Critical Radiant Flux - The lowest intensity of radiant heat that will cause a floor covering to propogate flame.
Crockfastness - Resistance of color to be transferred from carpet to a fabric.
Crocking - Removal of dye from carpet by rubbing.
Cross Section - Shape of a fiber across its cross section.
Cross Dyeing - Dyeing carpet with dyes, which have different affinity for different types of yarn.
Crushing - Matting of carpet fibers.
Crystallinity - Ordered structure of a polymer.
Cushioned Backed Carpet - Carpet, which has a cushion as a part of the backing.
Cut Loop Pile - Carpet with a combination of loop and cut pile tufts.
Cut Pile - Carpet with cut pile tufts.

D

Delamination - Separation of backing from carpet.
Delustering - Additives in fibers to reduce brightness.
Denier - A weight of yarn weighing one gram per 9000 meters.
Die - A device used to shape plastic into a sheet as it exits an extruder.
Die-land - Final parallel opening in a die.

Dimensional Stability - Ability of carpet to retain its original shape and size.
Drawing - Process of stretching fiber.
Dyeing - Method of coloring carpet.
Dye Lot - Carpet dyed at one time or carpet containing yarn dyed at one time.
Dyestuff - Material, which adds color to yarn by absorption into the fiber.

E

Emulsion - Suspension of fine particles in a liquid, generally water.
End - Individual yarn.
Equivalent Melt Index - Melt index value obtained from correlation of Melt Flow Rate.
Exhaust - To use up or take up.
Extrudate - Material exiting an extrusion die.
Extruder - Device with an internal rotating screw in a barrel, which transports material.
Extrusion - Process of forcing molten material through a die.

F

Face Weight - Total amount of face yarn in the carpet. Usually expressed in ounces per square yard.
Fading - Loss of color in carpet yarn.
Fastness - Property of dye to retain its color under different conditions.
Fiber - Basic material to form a textile product.
Fiber Shape - Cross section of fiber.
Fiber Size - Thickness of fiber filament.
Filament Yarn - Continuous lengths of fibers and packaged as a single yarn.
Finishing - Tufting and dyeing of carpet.
Flame Resistance - Describes a material that burns slowly.
Fluorochemical - Low surface energy chemical that contains fluorine.
Foot Traffic - A person walking on carpet.
Fungi - Mold.
Fuzzing - Fibers coming loose from carpet.

G

Gauge - Number of needles per inch across a tufting machine or stitches per inch.
Gram Stain - A stain which bacteria are classified by whether they retain or lose their primary stain.
Graphic Tufting Machine - Tufting machine capable of producing patterns.
Greige Good - Carpet undyed or unfinished.

H

Hand - How a carpet feels to human touch.
Heather - Slight multicolor appearance from mixing yarn or fiber.
Heat Setting - Setting a memory of twist into yarn using heat.
Hexamethylenediamine - Chemical used to produce nylon 6,6.
Hexapod Drum Test - Appearance retention test instrument using a hexapod in a drum.
Hollow Filament Fiber - Fibers with some hollow or interior void space.
Hot Melt - Thermoplastic compound, 100% solid at room temperature.
Hot Tack - Cohesive strength of a coating when molten.
Hydroxyl - Alcoholic group symbolized by (-O-H). It is the reactive group in polyols.

I

Isocyanate - Chemical containing (-N=C=O) group.
Inherent Viscosity - Viscosity of a polymer dissolved in a solvent.
Inoculum - Material containing microorganisms and used for inoculation.

L

Latex - Water emulsion of polymers or chemicals used to coat carpet.
Level Loop - All tufts in a loop form at the same height.
Lightfastness - Resistance of carpet to change color from sunlight.
Loop Pile - Carpet with all loop tufts.
Luster - Brightness of fiber or yarn.

M

Mastic - Highly filled, viscous, gap filling adhesive.
Matting - Crushing.
Medium - Substance used to provide nutrients for microorganism growth.
Melt Flow Rate - Melt index measured at 125°C.
Melt Strength - Ability of plastic to hold the die shape.
Melt Temperature - Temperature of plastic in an extruder.
Metering Zone - Section of extruder controlling rate of flow into die.
Methenamine Pill Test - Flammability test using a methenamine pill.
Microcrystalline Wax - Flexible and soft paraffin containing fewer straight chains.
Microorganism - Form of life of microscopic dimensions.
MDI - Abbreviation for diphenylmethane diisocyanate or diisocyantodiphenymethane.
Molecular Weight - Total number of the atomic weights of a molecule.

N

Nylon - Fiber used in carpet, usually nylon 6 or nylon 6,6.
Nylon 6 - Made from caprolactam.
Nylon 6,6 - Made from adipic acid and hexamethylene diamine.
Non-Woven Backing - Backing, which is not produced by weaving.

O

Olefin Fiber - Polypropylene fiber.
Open Time - Time between the applications of adhesive to substrate to form a bond.
Ozone Fading - Fading of carpet yarn from ozone.

P

Paraffin Wax - Low molecular weight, primarily straight chain hydrocarbon.
Pattern Match - Matching the pattern of carpet.
Pattern Streaks - Streaks, which occur in carpet from repeating direction.
pH - A symbol used to indicate acidity or alkalinity of a solution.
Piece or Beck Dyed - Carpet is dyed in a piece after tufting, but before coating.

Pigment - Insoluble material used to put color in a material.
Pile - Surface of carpet.
Pile Crush - Loss of pile thickness.
Pile Height - Length of yarn from the primary backing to the tip of the fiber.
Pile Reversal - Change in pile direction.
Pile Thickness - Thickness of actual yarn on the face of carpet.
Pile Weight - Weight of actual yarn on the face of carpet.
Pilling - Yarn coming loose from carpet surface.
Pill Test - Methenamine pill test.
Plush - Smooth level cut pile carpet.
Ply - Number of single yarn components twisted together into a yarn.
Polyester Fiber - Fibers containing polymeric ester of aromatic carboxylic acid.
Polymer - Large long chain chemical molecules.
Polypropylene Fiber - Fibers formed from chain growth polymerization of olefins.
Post-dyed - Carpet dyed in tufted form.
Pre-dyed - Carpet produced with pre-dyed yarn.
Pressure Sensitive Adhesive - Permanently tacky adhesive.
Primary Backing - Material into which tufts are placed to form carpet.
Primary Color - Red, blue, and green are the primary colors of light.
Printed Carpet - Carpet having printed patterns.
Pyrolysis - Decomposition of a compound caused by heat.

R

Radiant Panel Test - Flammability test using radiant energy to sustain flame.
Random Shear - Texture formed on carpet by shearing level or high loops.
Repeat - Pattern repeat in carpet.
Resilience - Ability of carpet to retain its texture.
Resin (Tackifier) - Low molecular polymers that impart adhesion and tack.
Ring and Ball Softening Point - Temperature measured in a ring and ball test.

S

Saxony - Thick cut pile carpet with yarns even across face.
SBR - Styrene butadiene rubber.

Screen Changer - Device which rapidly changes out screen packs.
Secondary Backing - Backing, which is attached to the back of carpet.
Selvage - Edge of carpet.
Shading - Color difference between one area of carpet face and another.
Shearing - Carpet manufacturing process using rotating blades to cut tips of yarn.
Shear Strength - Ability of adhesive to resist splitting.
Single Screw Extruder - Extruder that pumps plastic with one screw.
Singles Yarn - One yarn end.
Skein Dyed Yarn - Singles yarn, which has been dyed wound in a skein.
Smoke Chamber Test - Test method to measure smoke generation of carpet.
Soil Hiding - Ability of carpet to hide soil.
Soil Resistance - Ability of carpet to resist soiling.
Solution Dyed Yarn - Yarn, which has pigment added in extrusion process.
Space Dyed Yarn - Yarn which has been dyed at intermittent times.
Spun Yarn - Yarn made from short lengths of fiber.
Staple Yarn - Produced by spinning short lengths of cut filaments into yarn.
Static Control Test - Test to measure static discharge of carpet.
Static Charge - Electrostatic buildup in carpet.
Step Growth Polymerization -
Stitches per Inch - Numbers of tufts or stitches made by the tufting machine in the direction of the carpet.
Stocked Dyed - Yarns are dyed before spinning.
Suessen Heat Setting - Process using dry heat to set twisted yarn.
Superba Heat Setting - Process using steam and pressure to set twisted yarn.
Surface Area - Outer area at the surface of filament.
Surface Energy - Tendency of yarn to repel liquids.
Surfactant - A substance capable of reducing the surface tension of a liquid in which it is dissolved.
Synthetic Fiber - Man-made fiber.

T

Tack - Condition of adhesive when it wets the surface sufficiently to form a bond.
TDI - Abbreviation for toluene diisocyanate.
Telomerization - The formation of an addition oligomer.
Tensile Strength - Strength of fiber.

Texture - Surface characteristics of carpet.
Texture Retention - Crushing and matting resistance of carpet.
Texture Loop - Loop pile carpet with loops of different heights.
Texturizing - Process that adds texture to fiber.
Tip Definition - Visible single ends of cut yarn.
Tip Shearing - Shearing high loops of carpet.
Titanium Dioxide - Chemical used to de-luster fiber.
Tow - Continuous filaments in rope form.
Tuftbind - Force required to pull a tuft from carpet.
Tufted Carpet - Carpet formed by tufting.
Tufting - Carpet manufacturing process using needles to sew yarn into tufting primary.
Tufts per Square Inch - Number of tufts or stitches per square inch of carpet.
Twin Screw Extruder - Extruder using two screws to pump plastic.
Twist - Number of turns per inch of yarn.
Two-ply Yarn - Two single yarns twisted together.

V

Vetterman Drum Test - Instrument used to test appearance retention.
Vinyl - Common name for polyvinyl chloride or PVC.

W

Warp - Yarns used in the length direction of material.
Watermarking - Random shading or pile reversal.
Weft - Yarn used in the width direction of material.
Wool - Fiber made from sheep hair.
Woven Backing - Backing produced by a weaving process.

Y

Yarn - Continuous strand of fiber.
Yarn Count - Number used to describe the size of yarn.
Yarn Dyeing - Applying color to yarn.
Yarn Size - Weight of total filaments of yarn.
Yarn Weight - Amount of yarn used to produce carpet.

Bibliography

1. Commercial Carpet Digest, Infosource® (1990–1995)
2. Environmental Protection Agency.
3. International Wool Secretariat Publications.
4. Levy, S., and Carley, J., *Plastic Extrusion Technology Handbook*, 2nd Ed., Industrial Press Inc., NY (1989)
5. On Carpet, Infosource® (1986–1995)
6. Pelczar, M., Jr., *Microbiology*, McGraw-Hill Book Company, Inc., NY (1958)
7. Slade, P., *Handbook of Fiber Finish Technology*, Marcel Decker Inc., NY (1998)
8. Customer Processing Guidelines for Carpet Dyeing and Finishing, Allied Signal Inc. (1990)
9. Technical Data Sheets, American Association of Textile Chemist and Colorist (AATCC) Publications.
10. Technical Data Sheets, BASF Publications.
11. Technical Data Sheets, Diamond Shamrock literature.
12. Technical Data Sheets, Dow Chemical Company Publications.
13. Technical Data Sheets, E. I. DuPont de Nemours and Company Publications.
14. Technical Data Sheets, Equistar Chemicals Publications.

15. Technical Data Sheets, Hercules Inc. Publications.
16. Technical Data Sheets, Monsanto Company (now Solutia) Publications.
17. Technical Data Sheets, 3M (Minnesota Mining and Manufacturing Co.) Publications.
18. Textile Research Institute Publications.
19. Textile Research Journal Publication.
20. Woods, G., *The ICI Polyurethanes Book*, John Wiley & Sons, Chiester (1987)

Index

D

E

M

N

U

V

W

FSC
www.fsc.org
MIX
Papier | Fördert gute Waldnutzung
FSC® C083411

Kolibri 360 GmbH
Ferdinand-Jühlke-Straße 7
99095 Erfurt, Deutschland
produktsicherheit@kolibri360.de